Reinforced Concrete

Carbonation-induced corrosion of reinforced concrete is among the most widespread durability problems affecting service life of structures, and it underscores the need to develop quantitative engineering methods. This book promulgates the Natural Carbonation Prediction (NCP) model derived through experiments, data mining and analyses, along with mathematical formulations and refinements. It initially delves into fundamental concepts on corrosion of reinforced concrete, before employing approaches to service life design based on the applicative and practical stochastic methods supported by independent data.

Features:

- Focusses on corrosion due to carbonation of Portland cement-based concrete.
- Covers natural carbonation mechanisms, and probabilistic service life design through reliability analysis and Monte Carlo simulation.
- Presents a practical model for prediction of carbonation rate using a mathematical framework that incorporates key parameters for consideration in the engineering design process.
- Using data from worldwide sources, various case studies from different countries globally, are employed comparing measurements with carbonation predictions.
- Delves into carbonation modelling, originating from theoretical underpinnings based on Fick's laws of diffusion.

This book is aimed at graduate students, researchers and professionals in civil engineering, concrete technology and the built environment.

Reinforced Concrete
Natural Carbonation and Service Life Modeling

Stephen Ekolu

CRC Press
Taylor & Francis Group
Boca Raton London New York

CRC Press is an imprint of the
Taylor & Francis Group, an **informa** business

First edition published 2026
by CRC Press
2385 NW Executive Center Drive, Suite 320, Boca Raton FL 33431

and by CRC Press
4 Park Square, Milton Park, Abingdon, Oxon, OX14 4RN

CRC Press is an imprint of Taylor & Francis Group, LLC

© 2026 Stephen Ekolu

ISBN: 978-1-041-08350-4 (hbk)
ISBN: 978-1-041-08438-9 (pbk)
ISBN: 978-1-003-64539-9 (ebk)

DOI: 10.1201/9781003645399

Typeset in Times
by Apex CoVantage, LLC

Contents

Preface

The carbon dioxide (CO_2) driven climate change impacts observed worldwide, also highlight the significance of carbonation and corrosion of reinforced concrete structures. The already high atmospheric CO_2 concentration exceeding 400 ppm implies that the extent and severity of carbonation and reinforcement corrosion will correspondingly increase throughout the 21st century. This context underscores the important need to develop quantitative engineering methods of service life design. Despite the significant accumulation of knowledge and progress made on service life research over the past decades, this subject is not fully fledged for practice owing to complexity of the deterioration processes and the associated difficulty of developing practical engineering models.

It is also worth noting that while some corrosion prediction models have been proposed in the research literatures, the vast majority are experimental methods. Meanwhile, the handful of proposed practical models reported in the literatures, focus mainly on chloride-induced corrosion and much less on carbonation, understandably the former mechanism being more severe and damaging, especially under the cold winter climate of the Northern hemisphere. But most developing countries are located in the tropical and subtropical climate zones, wherein the stock volume of concrete infrastructures is amassing, and wherein carbonation is a bigger problem than chloride attack.

The Natural Carbonation Prediction (NCP) model promulgated in this book is the culmination of 12 years of the author's research comprising experiments, data mining and analyses, along with mathematical formulations and refinements. The model is uniquely based on natural carbonation and ambient environmental conditions, without in anywise involving laboratory-based accelerated carbonation method(s) of evaluating the deterioration process. The subject matter covered in this book initially delves into fundamental concepts on corrosion of reinforced concrete before employing approaches to service life design based on the applicative stochastic methods. The NCP model is fully described along with definition of its parameters. Independent data from worldwide sources are drawn and utilized for the model's validation and justification. The model is then applied to real-life concrete structures worldwide to evaluate its practical performance, wherein it is employed for service life prediction using reliability index analysis and Monte Carlo simulation technique.

This book provides useful knowledge for practicing civil engineers, the built environment professionals in the construction industry, structural and consulting engineers. Covered in the book are underlying concepts, theory and method(s) for service life design of reinforced concrete. Hence, the manuscript can be a useful resource for academic research and teaching of college or university students.

I unreservedly convey boundless and heartfelt gratitude to my PhD supervisors at University of Toronto and leading international academic experts on concrete durability, Professor R.D. Hooton and late Professor M.D.A. Thomas, for their impact that unwittingly laid essential foundational aspects of this book content.

I would like to acknowledge the contributions of my postgraduate and undergraduate students whose tenacious research on various experimental studies, impacted the NCP model development in different ways. Among them are Fitsum Solomon, Billy Ekolu Edamu, Maeteletsa Mashilo and Mashaba Mamabolo. Data sources from various literature sources worldwide were employed in the model's validation. While all publications are cited and fully referenced in various chapters of the book, sincere gratitude is extended generally to all authors whose data were employed in the NCP model at one stage or another. The author extends wholehearted appreciation to the Council for Scientific and Industrial Research (CSIR), South Africa for their original research data without which derivation of the NCP model would not have been possible. Another valuable category of contributors are the anonymous reviewers of the author's papers, whose insights led to improvement of the original model and its refinement to the final current version. Sincere thanks to all the reviewers that contributed their valuable insights of impact on the NCP model.

Ultimately, the author ascribes this manuscript to the enabling grace of God unmistakable during the intense personal challenges I encountered at early stages of the underlying scientific research. Strangely indeed during this period, remarkable creative insights and solutions emerged that form the core of this book.

Stephen Ekolu
Professor of Civil Engineering

Author biography

Stephen Ekolu is Professor of Civil Engineering and is rated among the World's Top 2% Scientists across all disciplines (Standford University/Elsevier; https://topresearcherslist.com/). He was formerly the Head of Civil Engineering Department at Nelson Mandela University (NMU). Before joining NMU, he was Associate Professor of Concrete Materials and Structures, and the Head of School of Civil Engineering and the Built Environment at University of Johannesburg. Previously he was a senior lecturer at School of Civil Engineering and the Built Environment, University of the Witwatersrand. He holds MSc (Eng) with distinction from University of Leeds, UK and PhD from University of Toronto, Canada. Prior to the postgraduate qualifications, he earned a BSc (Eng) in civil engineering from Makerere University, Uganda. Prof. Ekolu is an NRF rated researcher and PrEng professionally registered engineer with extensive experience in the industry, academic teaching and research. In total, he has over 27 years of civil engineering experience with more than 20 years in academia following seven (7) early career years of working in the industry.

Prof. Ekolu has published a combined number of over 240 peer reviewed journal papers, conference articles, books, research and technical reports. His teaching and research expertise focus is concrete materials and structures, and his concerted 12-year research on natural carbonation and service life prediction, culminated into development of the NCP model.

He is a reviewer (by invitation) for several leading international journals in the engineering field. He was Chair and Convenor of ICCMATS-1: International Conference on Construction Materials and Structures held on 24–26 November 2014 at Johannesburg, South Africa. He regularly provides consulting expertise and training in the industry.

Figures

Tables

Tables

Symbols and abbreviations

$[CO_2]$	carbon-dioxide concentration
μ	mean of
a	the unit amount of CO_2 that reacts with CH over the area (A) and across concrete thickness (dx)
c	cover depth
CBD	central business district
CEM I	ordinary Portland cement based on EN 197-1
cem	scalar quantity for the cement type
CH	calcium hydroxide
CI	computational intelligence
CoV	coefficient of variation
C_s	CO_2 concentration at the surface of concrete
CSH	calcium silicate hydrate
CV	coefficient of variation of error
C_x	CO_2 concentration at carbonation front in the interior of concrete
D	diffusion coefficient of concrete
$d_{c,t}$	time-dependent carbonation depth function of the NCP model
dC/dx	concentration gradient of $[CO_2]$ over the distance (dx) between the surface and interior of concrete
DDM	data-driven modelling
E	elastic modulus
e_c	environmental correction factor for CO_2 concentration
e_h	environmental correction factor for RH
e_s	environmental correction factor for sheltering
ESL	end-of-service life
e_t	environmental correction factor for temperature
F(P), P_f	failure probability
F(t)	time-dependent strength growth function
FA	fly ash
f_c	cube strength, f_{c28} or f_{cbn}
f_{c28}	28-day concrete strength
f_{cbn}	*in situ* concrete strength
f_{ck}	characteristic strength or strength grade
f_{cyl}	core or cylinder strength
g	exponent for the cement type
ggbs	ground granulated blast-furnace slag
J	diffusion flux, being the amount of CO_2 passing through a unit area (A) of concrete over a unit period of time (t)
J_{CO2}	the total amount (e.g., in grams) of CO_2 that diffuses into concrete across area (A) over time (t)
K_c	carbonation coefficient or rate
LL	limestone filler
MK	metakaolin

MV	mean value
N	Number of Monte Carlo simulation runs
NCP	natural carbonation prediction
NDT	non-destructive test
NN	neural networks
OPC	ordinary Portland cement
PV	predicted value
R	Resistance
q	temperature parameter for concrete
RH	relative humidity
RHC	rapid hardening Portland cement
RMS	root mean square of error
S	Loading
SCMs	supplementary cementitious materials
SF	silica fume
SG	slag
shl	sheltered
SLA	service life analysis
SLD	service life design
SLS	serviceability limit state
S_t	time-dependent carbonation depth function in the reliability equation
T	temperature
t	time
t_i	initiation time of deterioration in Tuutti's conceptual model
t_j	any time at $t > 0$
t_p	propagation time of deterioration in Tuutti's conceptual model
t_{sl}	service life
unshl	unsheltered
w/cm	water/cementitious ratio
β	reliability index
σ	standard deviation

1 Introduction

1.1 SOURCES OF STEEL CORROSION

Corrosion of steel reinforcement is the most widespread durability problem in concrete structures. Steel corrosion is responsible for major repairs and maintenance works required to extend the service life of reinforced concrete. It is well-documented that the two (2) main causes of steel corrosion in reinforced concrete, are chloride attack and carbonation. In the temperate climates, especially among the geographical regions of Europe and North America, along with some Asian countries in the Northern hemisphere, chloride attack is the dominant corrosion mechanism. Due to freezing conditions that occur during the winter season, ice or snowfall typically covers the land surface including roads, pavements and highway structures, thereby preventing their daily use. Typically, snowfall may occur overnight, covering walkways, roads and pavements right from residences or home doorsteps all the way to highways and offices, factories or other workplaces. Often, the resulting ice formed may last several weeks, if left undisturbed. Under these conditions, it is impossible for the public to utilize the infrastructures in a normal manner. In order to unblock roads and highways for vehicle use, de-icer salts which contain chlorides are typically sprayed to melt the ice before it is cleared. In turn, these chlorides get into concrete structures including pavements, bridges, culverts, parking structures, among others. The main natural source of chloride attack is seawater, which is relevant to coastal structures. Chloride attack is much more severe than carbonation. Hence in some regions, research on chloride corrosion attack overshadows that of carbonation.

Carbonation is a deterioration mechanism which occurs when CO_2 that is present in the atmosphere, penetrates into the concrete cover and comes into contact with steel reinforcement, causing its depassivation and corrosion. Being widespread globally, carbonation can occur in all inhabited environments including those under chloride exposure. Moreover, there is synergy between the two (2) mechanisms with carbonation playing the secondary role of promoting chloride attack by releasing chemically-bound chloride ions that are held within concrete. Of the two (2) damage processes, carbonation is the predominant mechanism occurring under tropical climate conditions at the inland areas, wherein it solely causes steel corrosion in reinforced concrete.

1.2 CLIMATE CHANGE IMPLICATIONS ON CONCRETE STRUCTURES

Anthropogenic climate change is one of the foremost challenges of the 21st century. Research has shown that since ancients times dating back to B.C. 800 000, cycles of *natural* climate change had maintained the $[CO_2]$ fluctuations of 180 to 280 ppm, as

DOI: 10.1201/9781003645399-1

shown in Figure 1.1a (Ekolu, 2023). But about 150 years from the onset of Industrial Revolution at around A.D. 1750, [CO$_2$] breached the 300-ppm level in A.D. 1900 and recently breached 400 ppm in A.D. 2014 (Figure 1.1b) (https://research.noaa. gov/2020/06/04/rise-of-carbon-dioxide-unabated/). It is projected that the [CO$_2$] could reach very high levels of 700 to 1000 ppm by A.D. 2100 (IPCC, 2007).

The implications of the ongoing rapid rise in [CO$_2$] on service life of concrete structures are expectedly significant. Investigations show that carbonation progression in concrete structures will increase significantly over the next decades. Talukdar et al. (2012) reported a carbonation increase of 45%, while Stewart et al. (2011) predicted carbonation depth increase of 25% to 36% over the 100-year period from A.D. 2000 and A.D. 2100. Ekolu's (2020) study reported a corresponding carbonation increase of up to 31% along with service life reductions by up to 24%. The countermeasure to the foregoing scenario is design of structures for climate resilience,

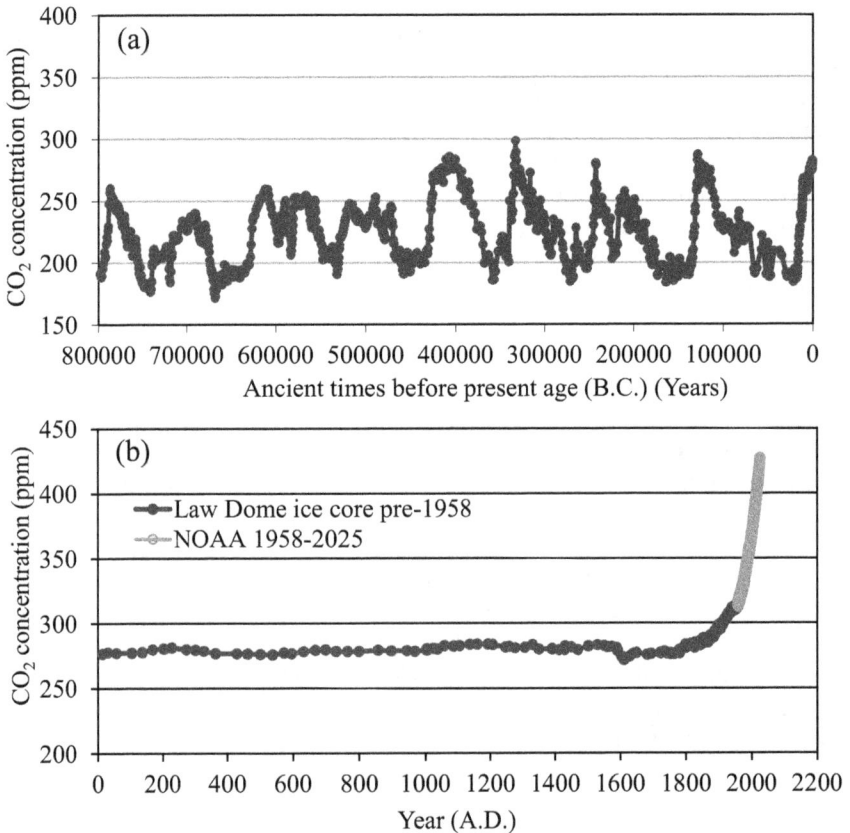

FIGURE 1.1 Historical CO$_2$ emissions depicting (a) natural climate change cycles in ancient B.C. times, (b) anthropogenic climate change since onset of Industrial Revolution in A.D. 1750.

Source: (data from Lüthi et al., 2008; NOAA, 2006, 2025).

which in turn requires employment of engineering analytical methods of quantifying service life.

The core aspect of structural design is load-bearing capacity analyses, with limited consideration of serviceability and durability. Presently, there is lack of practical service life design methods. Alexander (2018) attributes the lack of practical carbonation modelling techniques to complexity of the phenomenon, considering its wide range of variables.

1.3   CARBONATION AND CHLORIDE MECHANISMS IN GLOBAL CLIMATES

In the tropical and subtropical climate regions, carbonation is a much bigger problem than other attack mechanisms, as already mentioned earlier. In sub-Saharan Africa for example, chloride attack is of concern only alongside the seacoast of Atlantic and Indian Oceans, for a distance of up to 15 km inland from the coastline (SABS 0100–2, 1992; Alao et al., 2014). As expected, coastal cities are fewer and generally smaller than inland urban centres. The same scenario applies to the tropical areas of South America and Asia.

The spatial human occupancy of global land area based on climatic zones, is interesting. The area coverage by the tropical/subtropical region is 19.9% of the earth's surface, second to the temperate region which occupies 32.8%. Importantly, these two (2) regions are the most inhabited land areas occupied by about 70% of the world population. These regions are also the main locations of economic and industrial activities contributing to 81.2% of the world GDP (Mellinger et al., 1999). Notably, the largest stock volume and quantity of world infrastructures are located within these two (2) regions. The temperate climate is mostly found in the Northern hemisphere wherein are located western or industrialized countries including those in Europe, North America and Northern extremes of Asia. Most developing countries of Africa, Asia and South America are located in the tropics wherein dwell 40% of the world population, compared to 29.6% of the population living in the temperate climate region. Population densities of the two (2) regions are, however, similar. Due to their high population densities and industrial activities, these regions are highly urbanized, a significant factor that involves intensive utilization of economic infrastructures.

Tropical and subtropical climates have two (2) seasons of wet and dry weather, along with warm to hot temperatures ranging from 18 to 30°C. In contrast, the temperate climate has four (4) seasons comprising spring, autumn, winter and summer. This climate type is characterized by humid wet summer and severe winter with cold temperatures as low as below –20°C, along with snowfall. In the temperate climate wherein de-icer chloride salts are used to melt the ice after snowfall, chloride attack is a priority durability concern over carbonation. Moreover, chloride attack is more severe than carbonation attack, as discussed earlier. But in the tropical climate areas, the predominant corrosion mechanism is carbonation, except in coastal areas wherein chloride attack is the main damage attack process.

1.4   MATERIALS PARAMETERS FOR CARBONATION MODELLING

1.4.1   EARLY RESEARCH

Strength and permeability are the two (2) contending parameters advanced in the literatures as indicative concrete properties for use in carbonation modelling. Most early research studies on carbonation were conducted between 1960s and 1980's. During that period, the carbonation models proposed in the literatures were mostly strength-based models (Parrot, 1987; Kokubu and Nagatakis, 1989; de Fontenay, 1985). Parrot (1987) provides a comprehensive review on several of such early research studies done in the UK, Japan, Germany, Scandinavia and other countries. Evidently, there was an overwhelming preference to establish correlation between carbonation and strength, especially considering that the latter is a basic property employed in structural design.

As understanding of concrete technology and behaviour improved, it became clear that strength alone as considered in structural design, is not sufficient to cater for durability of structures. Research showed that almost all durability problems of concrete arise from external ingress of aggressive agents such as chlorides, CO_2, sulphate ions, moisture and other agents that penetrate into concrete. This realization attracted research interest on the permeability property of concrete. It is mainly from this point onwards that attempts to correlate carbonation with permeability or diffusion property, began to emerge. Indeed, it was not until the 1990's that the earliest permeability-based model was proposed by Parrott (1994), unlike strength-based models which had begun appearing 30 years earlier in the 1960's.

It can be seen from the foregoing, that there are two (2) categories of empirical carbonation models, being based on the indicative performance parameter used, which may be permeability/diffusion or strength. Extensive research studies have been done on both kinds of models as reported in the various literatures (Kokubu and Nagatakis, 1989; de Fontenay, 1985; Parrott, 1994; Xu et al., 1996; Papadakis et al., 1991; Duracrete, 2000; CEB-FIP, 2010; *fib* 34, 2006; Bob, 1999; Quillin, 2001; Sarja and Vesikari, 1996; Hakkinen, 1993; Ikotun and Ekolu, 2012; Ekolu, 2010, 2018). One of the arguments that is often advanced in favour of the permeability-based models is that this property directly depicts CO_2 diffusion into concrete, while strength is an indirect parameter. In the literatures, however, it is well-established that both strength and permeability are each strongly correlated with carbonation progression into concrete.

Studies (Hobbs, 1994; Khan and Lynsdale, 2002) have shown that a strong correlation exists between concrete strength and carbonation, much as such correlation similarly exists between permeability and carbonation. Interestingly, research data in the literatures, also show that there is strong correlation between carbonation and field-based or *in situ* concrete strength. However, only very few such data exist that correlate carbonation with field permeability measurements. The main reason for limited availability of field-based permeability data is the difficulty of conducting such field tests, whereas *in situ* strength measurements can be easily done either using the non-destructive (NDT) rebound hammer test or by extracting core samples for strength testing. Field permeability measurement equipment such as the Torrent permeameter, field air permeameter and other similar apparatus, are also relatively expensive. Moreover, field permeability tests can be difficult to set-up and cumbersome to conduct properly on site.

1.4.2 ADVANTAGES AND DISADVANTAGES

1.4.2.1 Method difficulty and problems

Since the second half of the 20th century, various direct and indirect methods including flow and penetration techniques, have been proposed for measurement of permeability. Moreover, attempts have been made to use different fluid types comprising air, gas and water (Hooton, 1988; Solomon and Ekolu, 2020). However, mostly indirect permeability test methods have attained some recognition through standardization, including the water absorption test (BSI 1881–122, 2011), rapid chloride permeability test (ASTM C1202, 2019), water penetration test (TS EN 12390–8, 2002) and triaxial water permeability test (CRD-C163, 1992).

The difficulties associated with permeability testing are highlighted in Hooton (1988). Among the important problems, is high variability of permeability results. It is also difficult or impossible to conduct direct permeability tests on high strength concretes. Moreover, permeability test methods typically do not represent the field boundary conditions. Collectively, these issues undermine the potential utilization of permeability as a modelling parameter. In contrast, strength testing does not have similar issues of major concern and adverse impact.

1.4.2.2 Variability of results

When employing engineering methods, consideration is given to variability of the results obtained. Standard deviation (σ) or the coefficient of variation (CoV) are among the main statistical parameters used to quantify variability of results determined using a given test method. It is desirable that variability of results obtained using any engineering procedure, is as low as possible. In engineering testing, however, this statistical quantity is affected by a combination of various factors including material property, experimental conditions and test procedure, among others. The high variability of permeability results is a well-known concern which undermines attempts to standardize permeability test methods. Extensive literatures have shown that variability of permeability test results is not only high but also occurs over a very wide range, with its CoV ranging from 30% to 150% as seen in Figure 1.2a (Hooton, 1988; Marsh, 1984; Hope and Malhotra, 1984; Day et al., 1985; Imamoto et al., 2006; Stanish et al., 2004). In contrast, concrete strength exhibits consistently moderate and acceptable variability, giving CoV values of typically between 15% to 30% (Figure 1.2b) (Shimizu et al., 2000; Ekolu, 2010; Sarja and Vesikari, 1996). Evidently, the variability of permeability measurements is at least an order of magnitude higher than that of compressive strength. High variability as observed for permeability testing, implies that repeatability and reproducibility of such results may not be meaningful owing to wide scatter of measured values. Accordingly, when considering an indicative performance parameter for carbonation modelling, it is evident from the foregoing that concrete strength is preferable over permeability.

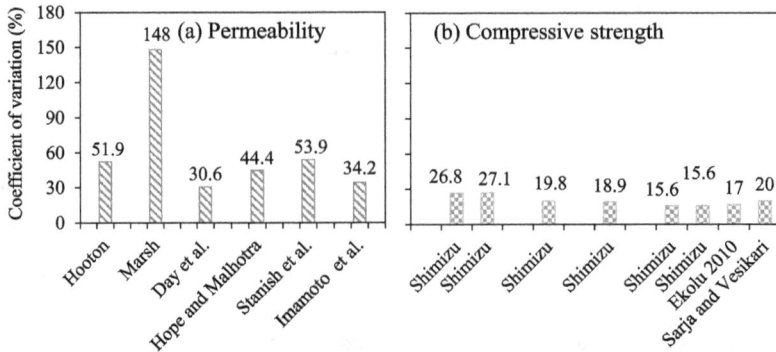

FIGURE 1.2 Variability of (a) permeability and (b) compressive strength results.

1.4.2.3 Cost of testing

It is desirable for test methods to possess the merits of simplicity and affordability, especially when considering their use in developing countries such as African nations and others. Proper permeability testing requires high quality technical training to ensure competence of the personnel. Employment of the permeability-based approach in carbonation modelling, would also involve acquisition of expensive test equipment along with the associated high labour cost of conducting the test. In South Africa for example, the cost of a single permeability test is about $100 compared to $11 for one (1) compressive strength test. Also, the test facilities for permeability measurement are not as widely available as those for compressive strength testing. Moreover, permeability test methods are mostly non-standardized techniques, unlike the concrete strength test which is a basic method that is standardized in all major national and international specifications.

Typically, concrete strength data of new structures are available from tests done during mix design or during the project construction. Moreover, the concrete strength grade that is used for structural design can also be employed for carbonation modelling and service life design. If *in situ* strength results are needed such as the case for existing structures, these data can be obtained simply by conducting NDT measurements using the rebound hammer method. The foregoing considerations show that the strength-based approach to carbonation modelling is more practically viable than that which employs the permeability property.

1.5 FUTURE OUTLOOK

Further to the discussion under Section 1.2, is another important countermeasure to the ongoing climate change impacts, being the development of low-carbon cement alternatives. It is well-established that Portland cement production is responsible for about 5% to 8% of global CO_2 emissions. Moreover by 2050, the global demand for cement is estimated to grow by 12% to 23% relative to its consumption in 2020, which implies intensified future rise in the CO_2 emissions based on current cement technologies (Belaid, 2022; Cheng et al., 2023). This forecast along with consideration of Portland cement's adverse environmental impacts, has led to the critical need to develop low-carbon cement alternatives that could contribute to the goal of achieving net-zero emissions by 2050 (Becken et al., 2024; Miller et al., 2021). Among the foremost of potential low-carbon systems undergoing intensive research are high-volume supplementary cementitious materials or SCMs for use as partial cement substitutes, limestone calcined clay cement or LC3 and geopolymer cement that could reduce up to 40% and 90% of the CO_2 emissions generated from Portland cement production, respectively. Being new material systems, intensive research is ongoing to fully understand their potential, performance characteristics and behaviours. The present version of the NCP model is based on standard Portland cements (Chapter 3), but it can in future be adapted or modified to cater for the evolving new low-carbon alternative cement systems.

1.6 OBJECTIVES

The objectives of this book are to introduce the NCP model and its application, along with provision of the foundational concepts on which the approach employed is based. Following the full description of the model and its parameters, focus is given to validation and demonstration of its veracity. It is then shown that the model gives realistic predictions, before shifting focus to its use for analysis and service life design, for both new and existing concrete structures.

Chapter 1 provides the background and historical context to the approach adopted in development of the NCP model, including the basis for choice of the core material parameter(s) employed. Consideration is also given to applicability of the model regarding its development with respect to method complexity, accuracy, implementation cost and practical aspects of relevance to developing countries.

Chapter 2 discusses the theory of steel corrosion in reinforced concrete, focusing on the carbonation mechanism. Tuutti's conceptual model of deterioration is used as the framework for service life design. The reliability analysis theory based on the applicative stochastic approach and the Monte Carlo simulation method, are introduced to provide an underpinning for practical implementation of the NCP model in later chapter(s).

Chapter 3 provides full description of the NCP model comprising its mathematical functions and a system of formulae. Employment of the model suits readers and users that have a background of an undergraduate engineering degree or its equivalent, covering relevant civil engineering courses including strength of materials, engineering mathematics, concrete materials and structures modules.

Chapter 4 presents an experimental justification of the model using worldwide data from independent research sources. For each data set, the model's predictions are compared with actual measured values of natural carbonation. Statistical error analysis is conducted to examine the model's prediction accuracy. To evaluate universal applicability of the model, the data employed are taken from worldwide sources covering the different human-inhabited global climates and geographical regions.

Chapter 5 focuses on application of the model to real-life existing concrete structures. Again, several independent data sources are utilized to demonstrate worldwide applicability of the model, along with its veracity. This chapter extends the principles already proven in Chapter 4, for application to the scenario of real-life engineering practice.

Chapter 6 concerns service life design using the NCP model. Stochastic methods comprising reliability index analysis and Monte Carlo simulation, are employed to extend application of the model from carbonation prediction as demonstrated in Chapter 5, to service life design in this chapter. The examples given in Chapters 5 and 6 illustrate the methodology of employing the model in engineering practice.

The first three (3) chapters of the book provide scientific and engineering principles that underpin the model's basis and form. The rest of the chapters underscore the performance of NCP model and its veracity when employed in engineering applications.

REFERENCES

Alao O.O., Alexander M. and Beushausen H. (2014) Understanding the influence of marine microclimates on the durability performance of RC structures. In: *Proceedings of the first international conference on construction materials and structures (ICCMATS)*, Johannesburg, South Africa, 1060–1067.

Alexander M.G. (2018) Durability and service life prediction for concrete structures – developments and challenges, *MATEC Web of Conferences*, 149, 01006.

ASTM C1202 (2019) *Standard test method for electrical indication of concrete's ability to resist chloride ion penetration*, ASTM International, West Conshohocken, PA.

Becken S., Miller G., Lee D.S. and Mackey B. (2024) The scientific basis of 'net zero emissions' and its diverging sociopolitical representation. *Science of the Total Environment*, (918), 170725, ISSN 0048-9697.

Belaid F. (2022) How does concrete and cement industry transformation contribute to mitigating climate change challenges?, *Resources, Conservation & Recycling Advances*, 15, 200084.

Bob C. (1999) Durability of concrete structures and specification. In: *Proceedings of the international conference on creating with concrete Dundee* (R.K. Dhir & M.J. McCarthy, Eds), 311–318.

BSI 1881–122 (2011) *Methods of testing concrete – method for determination of water absorption*, British Standard Institution (BSI), London.

CEB-FIP (2010) *Model code 2010, first complete draft – volume 2, fib bulletin 56*, International Federation for Structural Concrete (fib), Case Postale 88, CH-1015 Lausanne, Switzerland.

Cheng D., Reiner D.M., Yang F., Cui C., Meng J., Shan Y., Liu Y., Tao S. and Guan D. (2023) Projecting future carbon emissions from cement production in developing countries, *Nature Communications*, 14, 8213.

CRD-C163 (1992) *Standard test method for water permeability of concrete using triaxial cell*, US Federal Standards.

Day, R.L., Joshi, R.C., Langan, B.W. and Ward, M.A. (1985) Measurement of the permeability of concretes containing fly ash. In: *Proceedings 7th international ash symposium*, Orlando (2), 811–821.

de Fontenay C.L. (1985) Effect of concrete admixture, composition and exposure on carbonation in Bahrain, Deterioration and Repairs, Baharain, *Proc. VI*, 467–483.

Duracrete (2000) *General guidelines for durability design and redesign*, Report R15, in EU Brite-EuRam III project DuraCrete (BE95–1347): probabilistic performance based durability design of concrete structures.

Ekolu S.O. (2010) Model validation and characteristics of the service life of Johannesburg concrete structures. In: *Proceedings of the national symposium on concrete for a sustainable environment*, Concrete Society of Southern Africa, 3–4 August, Kempton Park, Johannesburg, Gauteng, 30–39.

Ekolu S.O. (2012) Model verification, refinement and testing on independent 10-year carbonation field data. In: *Proceedings of the third international conference on concrete repair, rehabilitation and retrofitting* (ICCRRR), 3–5, Sept. 2012, Cape Town, South Africa, 445–450.

Ekolu S.O. (2018) Model for practical prediction of natural carbonation in reinforced concrete: part 1-formulation, *Cement and Concrete Composites*, 86, 40–56.

Ekolu S.O. (2020) Implications of global CO_2 emissions on natural carbonation and service lifespan of concrete infrastructures – reliability analysis, *Cement and Concrete Composites*, 114, November 103744.

Ekolu S.O. (2023) Temperature-induced effect of climate change on natural carbonation of concrete structures, in the special issue on role of concrete and cement-based composites . . . and built environment, *ACI Materials Journal*, 120(1), https://doi.org/10.14359/51737335

fib 34 (2006) *Model code for service-life design, fib bulletin 34*, Federation International du Beton, Lausanne, 1st edition, 126p, ISBN: 978-2-88394-074-1

Hakkinen T. (1993) The influence of slag content on the microstructure, permeability and mechanical properties of concrete, part 2, technical and theoretical examinations, *Cement and Concrete Research*, 23, 518–530.

Hobbs D.W. (1994) Carbonation of concrete containing PFA, *Magazine of Concrete Research*, 46(166), 35–38.

Hooton R.D. (1988) What is needed in a permeability test for evaluation of concrete quality, MRS proceedings, *Materials Research Society*, 137, http://dx.doi.org/10.1557/PROC-137-141.

Hope B.B. and Malhotra V.M. (1984) The measurement of concrete permeability, *Canadian Journal of Civil Engineering*, 11, 287.

Ikotun J.O. and Ekolu S.O. (2012) Essential parameters for strength-based service life modeling of reinforced concrete structures – a review. In: *Proceedings of 3rd international conference on concrete repair, rehabilitation and retrofitting (ICCRRR)*, 3–5 September, Cape Town, South Africa, 433–438.

Imamoto k, Shimozawa K., Nagayama M., Yamasaki J. and Nimura S. (2006) Evaluation of air permeability of cover concrete by single chamber method. In: *31st conference on our world in concrete and structures*, 16–17 August, Singapore.

IPCC (2007) *Fourth assessment report of the Intergovernmental Panel on Climate Change (IPCC)*, Cambridge University Press, UK.

Khan M.I and Lynsdale C.J. (2002) Strength, permeability, and carbonation of high performance concrete, *Cement and Concrete Research*, 32(1), 123–131.

Kokubu M. and Nagatakis S. (1989) Carbonation of concrete with fly ash and corrosion of reinforcement in 20-years tests, *ACI. SP-114–14*, 315–329.

Lüthi D., Le Floch M., Bereiter B., Blunier T., Barnola J.M., Siegenthaler U., Raynaud D., Jouzel J., Fischer H., Kawamura K. and Stocker T.F. (2008) High-resolution carbon dioxide concentration record 650,000–800,000 years before present, *Nature*, 453, 379–382, Supplementary Information, https://doi.org/10.1038/nature06949

Marsh B.K. (1984) Relationships between engineering properties and microstructural characteristics of hardened cement paste containing pulverized fuel ash as a partial cement replacement, *PhD Thesis*, The Hatfield polytechnic, UK.

Mellinger A.D., Sachs J.D. and Gallup J.L. (1999) *Climate, water navigability, and economic development*, Center for International Development at Harvard University, CID Working Paper No. 24 September, 30p.

Miller S.A., Habert G., Myers R.J. and Harvey J.T. (2021) Achieving net zero greenhouse gas emissions in the cement industry via value chain mitigation strategies, *One Earth*, 4(10), 1398–1411.

NOAA (2006) *Law Dome ice core 2000-year CO₂, CH₄, and N₂O data*, National Oceanic and Atmospheric Administration (NOAA), https://www.ncei.noaa.gov/pub/data/paleo/icecore/antarctica/law/law2006-co2-noaa.txt

NOAA (2025) *Mauna loa CO₂ monthly mean data (text) or (CSV)*, National Oceanic and Atmospheric Administration (NOAA), https://gml.noaa.gov/webdata/ccgg/trends/co2/co2_mm_mlo.txt

Papadakis V.G., Vayenas C.G. and Fardis M.N. (1991) Fundamental modelling and experimental investigation of concrete carbonation, *ACI Materials Journal*, 4(88), 363–373.

Parrot L.J. (1987) A review of carbonation in reinforced concrete, *Cement and Concrete Association*, BRE, 42p.

Parrott P.J. (1994) Design for avoiding damage due to carbonation-induced corrosion, durability of concrete. In: *3rd international conference*, Nice, France, 283–298.

Quillin K. (2001) *Modelling degradation processes affecting concrete*, BRE Centre of Concrete Construction, CRC Ltd 151 Rosebery Avenue, London, ECIR 4GB, ISBN 1860815316.

SABS 0100–2 (1992) *Code of practice for the structural use of concrete, part 2: materials*, South African Bureau of Standards, Pretoria.

Sarja A. and Vesikari E. (1996) *Durability design of concrete structures*, RILEM Report 14 (A. Sarja & E. Vesikari, Eds), E & FN Spon, UK, 165p.

Shimizu Y, Hirosawa M. and Zhou J. (2000) Statistical analysis of concrete strength in existing reinforced concrete buildings in Japan. In: *12th world conference on earthquake engineering (WCEEE)*, Auckland, New Zealand, 30 January–4 February 2000, 8p. Built [1] before 1961, [2] 1961–65, [3] 1966–70, [4] 1971–75, [5] 1976–80, [6] after 1980.

Solomon F. and Ekolu S.O. (2020) Comparison of various permeability methods applied upon clay concretes–statistical evaluation, *Journal of Testing and Evaluation*, July, 48(4), JTE20160546.

Stanish K., Alexander M.G. and Ballim Y. (2004) *Durability index interlaboratory test results–statistical analysis of variance*, Research Report, Departments of Civil Engineering, University of Cape Town/University of the Witwatersrand, 87p.

Stewart M.G., Wang X. and Nguyen M.N. (2011) Climate change impact and risks of concrete infrastructure deterioration, *Engineering Structures*, 33, 1326–1337.

Talukdar S., Banthia N., Grace J.R. and Cohen S. (2012) Carbonation in concrete infrastructure in the context of global climate change: part 2, Canadian urban simulations, *Cement and Concrete Composites*, 34, 931–935.

TS EN 12390–8 (2002) *Testing hardened concrete – part 8: depth of penetration of water under pressure*, Institute of Turkish Standards, Ankara.

Xu A., Rodhe M. and Chandra S. (1996) Influence of alkali on carbonation of concrete. In: *Durability of building materials and components*, 7, vol. 1, C. Sjostrom (Ed.), Spon, 596–604.

2 Steel corrosion and service life

2.1 CARBONATION MECHANISM

This chapter focusses on understanding of the carbonation mechanism and its theoretical underpinnings for the purpose of service life design. Carbonation is one (1) of the two (2) mechanisms responsible for steel corrosion in reinforced concrete structures, the other being chloride attack. The two (2) mechanisms are typically treated separately owing to major differences in their attack processes. As already highlighted in Chapter 1, although chloride attack is more severe than carbonation, the former occurs in particular geographical locations comprising the coastal regions and cold temperate climates of the Northern hemisphere, while carbonation can occur at any region on earth globally. In the tropical and arid climates, carbonation is the main mechanism responsible for steel corrosion of concrete structures. Climate change, which is associated with the ongoing rise of anthropogenic CO_2 in the earth's atmosphere, has also brought to limelight the carbonation attack mechanism and its implications on resilience of concrete structures.

Upon diffusion of CO_2 into concrete as illustrated in Figure 2.1, chemical reactions ensue that lead to steel corrosion. Concrete is highly alkaline with a typical pH level of about 12.6 to 13, being sustained by mass presence of calcium hydroxide (CH) formed as a by-product of cement hydration (Neville, 1996; Fulton's, 2001). At such high alkalinity, a protective passive film layer forms at steel surface and prevents corrosion. During exposure of reinforced concrete structures to the surrounding environment, CO_2 penetrates into concrete and reacts with CH, in turn leading to decrease in alkalinity to pH of about nine (9). At such low pH, the protective passive film at the steel surface breaks down, exposing the reinforcement to corrosion.

As already mentioned, once CO_2 has penetrated into concrete, it reacts with CH, pore solution alkalis and calcium silicate hydrate (CSH) or tobermorite, reducing pH and causing steel depassivation. The predominant carbonation reaction is that of CO_2 with CH, to form $CaCO_3$ or calcite as shown in Equation (2.1a). A minor carbonation reaction also occurs between CO_2 and CSH to form carbonate-silicate, as given in Equation (2.1b)

$$Ca(OH)_2 + CO_2 \rightarrow CaCO_3 + H_2O \tag{2.1a}$$

$$3CaO \cdot 2SiO_2 \cdot 3H_2O + 3CO_2 \rightarrow 3CaCO_3 \cdot 2SiO_2 \cdot 3H_2O \tag{2.1b}$$

The presence and contribution of other pore alkalis consisting of NaOH and KOH, is typically minor, since CH is predominant within groundmass of the cementitious

DOI: 10.1201/9781003645399-2

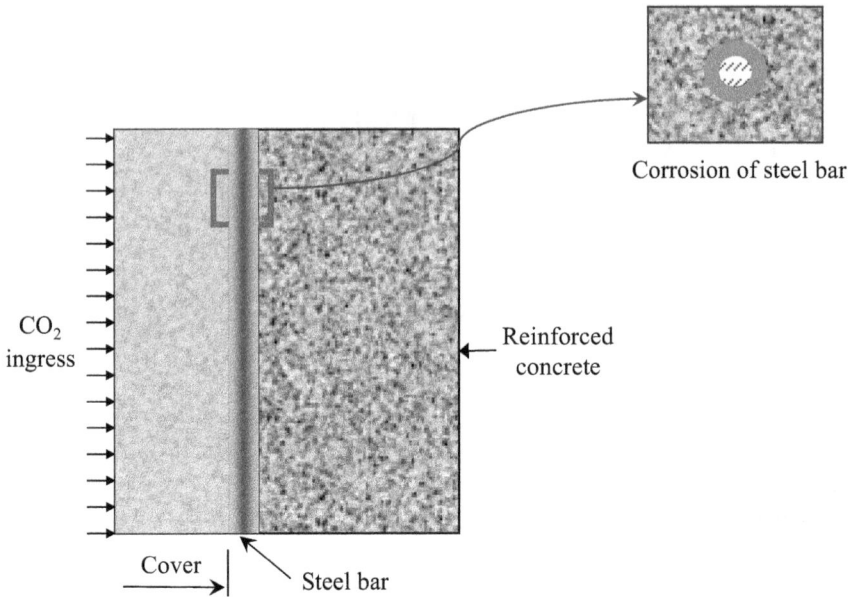

FIGURE 2.1   Carbon-dioxide penetration into reinforced concrete to induce steel corrosion.

matrix. It may also be noted that $CaCO_3$ (calcite) which is formed during the carbonation reaction (Eq. 2.1a), has very low solubility and consequently precipitates in concrete pores (Sagues et al., 1997). The formation of $CaCO_3$ within concrete pores, has interesting implications on concrete properties comprising the reduction in permeability and increase in compressive strength. Another important observation from this reaction process is formation of water within vicinity of steel reinforcement, which is one of the necessary support conditions for corrosion. Altogether, conditions for the corrosion process involve not only the depassivation of steel, but an adequate presence of moisture, oxygen availability and conductance of ions. Mobility of ions occurs through pore fluid movement while oxygen availability arises from ingress concurrently with CO_2 penetration, in addition to some dissolved oxygen that may also be present in pore water.

It is evident from the foregoing set of conditions that, the trigger to corrosion of steel is depassivation due to pH reduction upon ingress of CO_2 into the full cover depth, to reach the level of steel reinforcement. The other conditions comprising the presence of moisture and oxygen, along with mobility of ions, are readily present within the concrete pore network.

Corrosion of steel reinforcement is a three-step reaction process given in Equations (2.2a) to (2.2c) (Broomfield, 2007). The rust formed may change form, typically consisting of ferrous hydroxide (brown rust) or ferric hydroxide (black rust). More importantly, these corrosion products occupy a larger volume of about two (2) to six (6) times that of the reactants, depending on the type of final reaction product.

As more volume of the corrosion product is formed, expansive stress increases. Consequently, carbonation-induced corrosion leads to cracking, spalling and delamination of concrete along with reduction in cross-sectional area of the steel bar.

Testing of concrete carbonation is done using the phenolphthalein indicator solution which is prepared by mixing 1 gram of phenolphthalein powder with 70 mls of ethanol and 30 mls of distilled water (EN 14630, 2006; RILEM CPC-18, 1988). Typically, a concrete core or cube is split onsite or in the laboratory, then immediately sprayed with phenolphthalein indicator solution. Colour of the sprayed surface changes to pink at inner non-carbonated concrete, but remains clear at the outer carbonated depth which can then be measured.

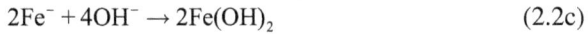

$$2Fe^{2+} \rightarrow 2Fe_2 + 4e^- \tag{2.2a}$$

$$4e^- + O_2 + 2H_2O \rightarrow 4OH^- \tag{2.2b}$$

$$2Fe^- + 4OH^- \rightarrow 2Fe(OH)_2 \tag{2.2c}$$

2.2 DIFFUSION THEORY AND CARBONATION

Ingress of ions or agents into porous media under concentration gradient, is universally based on the fundamental physics theory described by Fick's laws of diffusion. Given in Equation (2.3a) is Fick's first law from which the carbonation equation can be derived, as subsequently done.

$$J = D \cdot \frac{dC}{dx} \tag{2.3a}$$

where

   J is diffusion flux, which is the amount of ionic mass agent (CO_2 in this case) passing through unit area (A) of porous medium over a unit period of time (t). dC/dx is the concentration gradient of CO_2 over the distance, dx across the surface concrete and interior concrete. Concentration gradient is the driving force of ingress. D is diffusion coefficient of the material system (concrete in this case) and is the parameter representing connectivity of the pore network.

   From Fick's first law (Eq. 2.3a), the amount of CO_2 diffusing into concrete must be equal to the amount that participates in the carbonation reaction within concrete. Since the CO_2 gas at the carbonation front reacts with CH during the carbonation reaction, it follows that its concentration at the concrete surface exposure (C_s) is not the same as its concentration at the carbonation front (C_x) that is, $C_s \neq C_x$. Therefore, across any thickness, dx, between two (2) points of diffusion, Fick's first law can be written as given in Equation (2.3b). Accordingly, the total amount, J_{CO_2} (e.g., in grams) of CO_2 that diffuses into concrete across the area (A) over the time (t), is given by Equation (2.3c).

$$J = D \cdot \frac{C_s - C_x}{dx} \text{ per unit area } (A) \text{ of diffusion over time } (t) \qquad (2.3b)$$

$$J_{CO_2} = D \cdot A \cdot \frac{C_s - C_x}{dx} \cdot t \qquad (2.3c)$$

CO_2 that penetrates into concrete reacts with CH to form calcite. If the unit amount of CO_2 that reacts with CH is taken as a (in grams) over area A and across concrete thickness dx, then the total amount of CO_2 consumed in the carbonation reaction is given by Equation (2.3d).

$$J_{cbn} = a.A.dx \qquad (2.3d)$$

For equilibrium, $J_{CO_2} = J_{cbn}$. Hence by equating Equation (2.3c) with Equation (2.3d) and integrating both sides, Equation (2.3e) is obtained. Since parameters within the brackets of Equation (2.3e) are constant at any given time, the expression can be simplified to the more familiar Equation (2.3f) (Kropp, 1995).

$$\text{Carbonated concrete thickness (x) is } d_c = \left(\frac{2D(C_s - C_x)}{a} \right)^{1/2} \cdot \sqrt{t} \qquad (2.3e)$$

$$d_c = K_c \sqrt{t} \qquad (2.3f)$$

where K_c is the constant carbonation rate or carbonation coefficient of concrete.

But K_c is not an absolute constant. Rather, the carbonation coefficient depends on concrete ingredients, mixture and environmental exposure conditions. Given the infinite variety of concrete mixtures and the associated material systems, along with unpredictable fluctuations of environmental exposure conditions at any given geographical location, K_c is evidently a complex parameter that is difficult to realistically predict or estimate.

2.3 HISTORICAL EVOLVEMENT OF CARBONATION RESEARCH

2.3.1 PAST DEVELOPMENTS ON THE SUBJECT-THEORY-METHOD TRIAD

Carbonation is a natural phenomenon like most other mechanisms of concrete such as creep and drying shrinkage, chloride diffusion, permeability, water absorption and others which occur universally in porous material systems. These phenomena can be described using the fundamental laws of physics, chemistry, mathematics and mechanics, which govern the natural processes. A scientific theory is an abstraction of observed facts typically configured as a universal fundamental law. The diffusion of gases and fluids through porous media, is governed by the theoretical Fick's laws of diffusion that employ the scientific knowledge of physics and mathematical functions to express observable physical manifestation or effects of a given phenomenon. Indeed, the fundamental Fick's laws of diffusion govern the ingress of oxygen, CO_2, chlorides or moisture, into concrete. Accordingly, this fundamental theory provides the scientific principle of describing the phenomena and formulating the relevant prediction methodologies. Once a phenomenon is a well-established observation and its theory is understood, development of engineering methods is necessary for use to evaluate and analyse the process in practice. It is stated that both theory and methods emerge from the phenomenon (Klaus, 1965; Eder, 2014), implying that none of the three (3) knowledge components can meaningfully exist without the others.

Figure 2.2 shows interconnections between the different elements forming the subject-theory-method triad. The subsequent section provides some reflections that highlight the carbonation research trajectory over the past decades. The carbonation review by Parrott (1987) is perhaps the earliest major bibliography that attempted to compile the evolvement of carbonation research. It is evident that interest in carbonation emerged around late 1950s to mid-1960's, presumably being the period when the earliest existing reinforced concrete structures, began to majorly exhibit corrosion damage effects. At that point, carbonation of concrete had emerged as a subject attributed to natural behaviour (Figure 2.2) and so its effects attracted research interest. Among the notable earliest studies on concrete carbonation were those by Leber and Blakey (1956), Verbeck (1958), Hunt et al. (1958), Cole and Kroone (1960). Since then, research studies have covered the full spectrum of the triad on various knowledge aspects at various stages subsequently discussed (Parrott, 1987).

2.3.1.1 Carbonation reactions

Early research focused on understanding the carbonation reactions and identifying the sources of atmospheric CO_2 mainly comprising: the natural emitters including

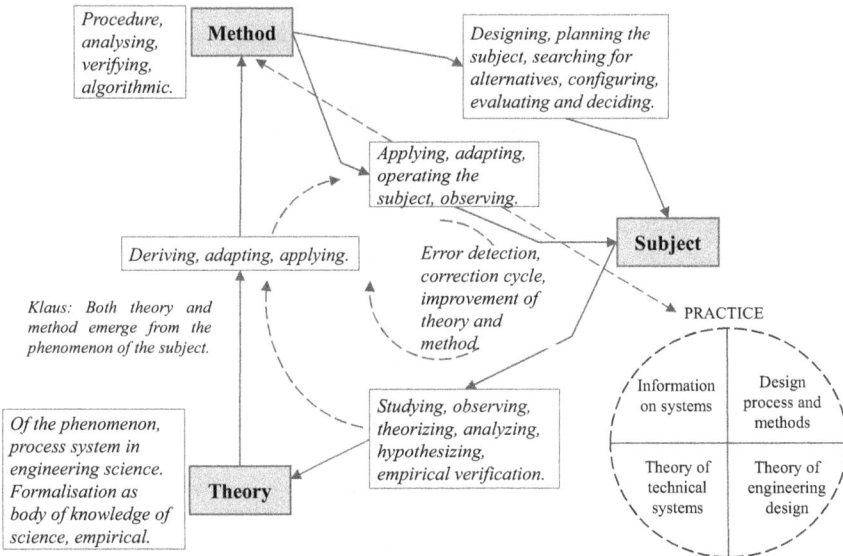

FIGURE 2.2 The subject-theory-method triad of phenomena.

Source: (adapted from Eder, 2014).

plants or vegetation, and artificial emitters being fossil fuels and industries. This basic research revealed key understanding of CO_2 chemical reactions with cement hydrates comprising CH, CSH and aluminate hydrates (Eqs. 2.1a, 2.1b), along with development of knowledge regarding the influence of carbonation reactions on pore fluid alkalinity, leading to a drop of pH from about 12.6 in non-carbonated concrete to below 9.0 after carbonation.

2.3.1.2 Measurement of carbonation

Understanding of the carbonation effect on pH of concrete, underpinned the development of the phenolphthalein indicator test. This technique is widely employed as the conventional method of identifying and measuring carbonation progression, owing to its simplicity, convenience and reproducibility of results. Alternative measurement methods have been explored including use of other different indicator solution types, chemical extraction techniques and analytical methods comprising X-ray diffraction, thermal analysis and optical microscopy. But none of the foregoing techniques surpasses use of the standard phenolphthalein indicator test method.

2.3.1.3 Accelerated carbonation testing

Accelerated carbonation involves exposing concrete samples to high CO_2 concentration under laboratory-controlled conditions. This approach allows rapid evaluation of concretes subjected to carbonation. The natural CO_2 concentration is presently in the range of 0.04% breached in 2014 (*https://research.noaa.gov*), whereas concentrations of 5% to 20% CO_2 are commonly used for accelerated carbonation in the laboratory. In fact, the accelerated carbonation method has been predominantly employed since around the 1970s. For example, Hamada (1969) reported the correlation of carbonation depths for concretes subjected to accelerated carbonation at the different concentrations of 0.045% CO_2 and 15% CO_2. As a consequence of employing the laboratory-based accelerated carbonation method, intensified research could be done for theoretical understanding on the effects of different variables, while avoiding the otherwise prohibitive constraint of prolonged time required to attain natural carbonation of samples.

2.3.1.4 Understanding of factors affecting carbonation

This category of investigation falls within the theory-subject segment of the triad (Figure 2.2), an aspect that is essential for evaluation of the phenomenon in engineering practice. Research into the factors influencing carbonation, emerged mostly in the 1970s and 1980s following understanding of the carbonation theory, along with development of the test methods discussed in Sections 2.3.1.2/3 above. Various important factors affecting concrete carbonation have been investigated over the past several decades. These factors comprise concrete mixture parameters including water/cementitious ratio (w/cm), cement content and cement extenders or supplementary cementitious materials (SCMs), among others. Prior to 1990s, the common extenders or SCMs used in the cement industry were ground granulated blast-furnace slag (GGBS) and fly ash (FA). Also in the early research, compressive strength was the main physical property considered for correlation with carbonation progression, more so since strength is the basic property used in structural design. Effects of various processing factors and concrete ingredients have been evaluated including those due to compaction, bleeding, curing, aggregate types and chemical admixtures. Indeed, in the 1980s, a great deal of intense research work was particularly done to investigate the effects of curing on carbonation.

Investigations to evaluate the effects of environmental factors, intensified between 1950s and 1980s, especially after development of the accelerated carbonation method and the phenolphthalein indicator test. Among the important environmental exposure conditions of interest were CO_2 concentration, relative humidity (RH) and temperature. It should be added that research on the foregoing is still ongoing for improved understanding of their effects on carbonation in the context of modern concrete technology and future infrastructure.

2.3.1.5   Effects of carbonation on concrete properties

Early understanding of carbonation effects on concrete properties was mostly based on research developed from 1960s to 1980s, alongside the investigations described in Section 2.3.1.4. Carbonation effects were found to be largely beneficial to mechanical properties including reduction of porosity, increase in compressive strength, increase in elastic modulus, increase in surface hardness and reduction of permeability (Parrott, 1987). However, carbonation progression is confined to a thin surface layer of the concrete cover, hence its beneficial effect on bulk properties is minimal. In contrast, the synergy of carbonation with other damage mechanisms such as chloride attack, exacerbates adverse effects.

2.3.1.6   Carbonation-induced steel corrosion

Steel corrosion of reinforced concrete is the most important problem responsible for scientific interest on carbonation. The resulting pH reduction owing to the carbonation reaction, leads to corrosion of the embedded steel reinforcement in concrete structures, in turn causing the associated structural effects comprising cracking of the concrete cover, spalling, delamination and loss of steel area. Durability of concrete structures has been a major focus of engineering research, since around the last quarter of the 20th century. In the earlier decades, interest in carbonation was largely overshadowed by research on chloride attack. Hence, the volume of available literatures on carbonation-induced steel corrosion is not as extensive as that of the research done on chloride attack. But since the first quarter of the 21st century, more research has continued to emerge focused on carbonation-induced steel corrosion. Presently, large stock volume of concrete infrastructure continues to amass in the tropical climate regions wherein highly populated cities in developing countries are located. In these regions within the inland areas, carbonation is generally a bigger problem than chloride attack.

2.3.1.7   Carbonation prediction

The need to mitigate adverse effects of carbonation leading to steel corrosion, necessitated the development of durability design specifications, especially those focused on the material system and cover depth, both of which provide protection to steel reinforcement against carbonation attack. Meanwhile, attempts to develop models for carbonation prediction have been made since the 1960s, and these efforts are ongoing. A plethora of experimental research models have been proposed during the past over six (6) decades. However, practical models for service life design have not been fully promulgated, owing to complexity of factors influencing carbonation. The earlier experimental models proposed in the 1960s to 1980s, were mostly simple empirical equations. Towards the end of the 20th century to the present time, more complex models with practical potential have been proposed (Ekolu, 2018). Obviously, the more recent models are methods that have emerged based on past decades of collective carbonation research findings.

2.3.1.8 Natural carbonation under outdoor conditions

In the earlier years since 1960s, outdoor-related data comprised *in situ* tests done on existing concrete structures, but lacked records of other important parameters. During that past period of time, the important factors affecting carbonation were not fully understood yet, hence such data were not only limited but lacked historical record of environmental parameters needed for full interpretation and utilization of findings. More useful data of *in situ* measurements began to emerge towards end of the 21st century, wherein were reported some essential environmental parameters for carbonation modelling.

Since the 1990s, there has been increased focus on the experiments that are conducted under natural carbonation outdoors. Among the earliest of such natural carbonation investigations was the 10-year CSIR (1999) study. Such research studies continue to increasingly generate valuable data necessary for development of potential practical models for engineering applications.

2.3.2 Discussion of the subject-theory-method triad

It is evident that evolvement in scientific understanding of the carbonation attack mechanism, as earmarked by the eight (8) aspects discussed in Section 2.3.1, has been a continuous spiral interplay of the *subject-theory-method* triad shown in Figure 2.2. Studies on the carbonation subject are ongoing, ranging from factors affecting the mechanism to its effects on concrete performance. A great deal of carbonation-related knowledge has been accumulated comprising the effects of concrete ingredients and mixture proportions, along with effects of concrete properties such as strength, porosity or permeability. However, modern concrete technology continues to evolve with emergence of new materials such as the expanding range of different pozzolan types, special aggregate types including recycled aggregates with a futuristic focus on sustainability, and special concretes such as self-compacting concretes, sprayed concretes, fibre-reinforced concretes, among others. Coatings and repair materials are so diverse and widely employed in the modern industry, hence their effects on carbonation are of interest especially with respect to practical carbonation modelling. Possible emergence of geopolymer binders as an alternative cementitious system to Portland cement, could bring about a significant shift of corrosion research focus regarding durability of reinforced concrete.

With climate change impacts being the crucial core focus of environmental issues globally in the 21st century, the impact of anthropogenic CO_2 rise on durability and corrosion of reinforced concrete structures will be a major concern for the next several decades. Durability specifications will require updating for use to design the climate resilience of concrete structures. Meanwhile for practical service life design, it is necessary that practical carbonation prediction model(s) are developed.

Fick's laws of diffusion provide the underlying fundamental theory describing concrete carbonation (Section 2.2). But factors that influence the carbonation reaction mechanism are also governed by different mathematical laws. For example, the response of concrete to temperature influence is governed by the Arrhenius equation, regardless of the reaction mechanism. Similarly, the influence of other material and environmental factors such as RH, CO_2 concentration, sheltering and SCMs or extenders, are each governed by different mathematical functions, that have to be discovered and properly configured for carbonation modelling.

Development of the early methods comprising accelerated carbonation test procedure and the phenolphthalein indicator test, particularly played an important role of enabling intensive research studies to be conducted relatively fast, in turn producing results that could be compared and contrasted to reach conventional findings. Perhaps the most challenging research aspect that started as early as 1960s and presently continues is that of carbonation modelling. It is worth noting that research on practical modelling of carbonation has been ongoing for decades with achievement of incremental developments, but still of limited use hitherto.

2.4 SERVICE LIFE CONCEPT

Corrosion of steel reinforcement is considered the most widespread and greatest threat to durability of reinforced concrete structures (Alexander, 2018). Once corrosion ensues, damage is progressive and when left unarrested, it can attain severity levels that lead to serviceability failure and can ultimately cause structural damage.

The concept of service life is represented schematically using the widely recognized Tuutti's (1982) conceptual model comprising two (2) stages namely: corrosion initiation and corrosion propagation. During the initiation time (t_i), aggressive agents especially CO_2/Cl^- penetrate into the concrete cover. Upon reaching the level of steel reinforcement after ingressing through the full cover depth, corrosion ensues during the propagation period (t_p), resulting in progressive exhibition of distress features of increasing severity from cracking to spalling and delamination, along with loss of steel area that can significantly reduce structural capacity. By definition therefore, service life is given by Equation (2.4).

$$t_{sl} = t_i + t_p \qquad (2.4)$$

where t_{sl} is service life, t_i is the initiation time period and t_p is the propagation period of deterioration.

The end-of-service life (ESL) typically lies anywhere from t_i to t_p depending on the attack scenario. Determination of t_i or t_p at design stage, requires employment of prediction models. While a great deal of research has over the decades been done on modelling of corrosion propagation, it is largely experimental and mostly based on chloride attack (Ikotun, 2017; Liang et al., 2012). In principle, ESL is typically specified as a criterion judged to be appropriate to safeguard the serviceability state of the structure. The two (2) points that are usually regarded to be the serviceability limit state (SLS) for ESL are either t_i or t_p at the point of cracking. The ESL at t_i is typically used for chloride-induced corrosion, since this attack is relatively severe. Hence ESL is deemed to occur once the critical chloride concentration has reached the level of steel reinforcement. Carbonation is relatively less severe than chloride attack. As such, the ESL under carbonation-induced corrosion can be less stringent than that of chloride attack. Cracking of the concrete cover is usually considered appropriate for use as ESL for carbonation-induced corrosion (Sarja and Vesikari, 1996). More detailed discussion of ESL criteria is provided in Section 6.3.

2.5 STOCHASTIC SERVICE LIFE DESIGN

2.5.1 THEORY OF FAILURE PROBABILITY

The general safety principle of engineering design is provision of resistance (R) that is greater than the applied loading (S) acting upon the structure. The two (2) parameters (R, S) can be any quantities of any units that cancel out each other. Both R and S are usually variables at any given time and are therefore more realistically represented as a normal distribution curve or other forms of distribution functions such as the log-normal or Weibull distribution. For example, the loading acting upon an office building varies throughout the day, week or month. As employees report to work in an office building on a particular working day, the structure is subjected to peak loading in the morning at say from 08.00 to 10.00 hours. This loading may reduce in the afternoon as some staff members get out to conduct assigned work duties outside office. This scenario of varying loading continues throughout the day until after 17.00 hours when everyone vacates office at end of the day's work, leaving the structure at its least loading throughout the night. On holidays and weekends, loading is also at its lowest. If the foregoing loading routines are recorded daily over a prolonged period of time throughout the month or year, the pattern obtained would likely depict the normal distribution curve.

Safety in engineering design is achieved through use of safety factors. In structural design for example, the anticipated loading acting upon the structure is calculated, then safety factors are applied to obtain higher design loads. The structure is then designed for higher capacity based on design loads which ensures that R > S, also written as given in Equation (2.5) (Sarja and Vesikari, 1996).

$$R - S \geq 0 \qquad (2.5)$$

Failure occurs once resistance is lower than loading. The same principle applies to durability design, wherein the loading S is the measure of deterioration ingress attack such as CO_2/Cl^- penetration, while R is the resistance to ingress of attack agents. While load-induced structural failure can be a single event such as an abrupt collapse due to impact or due to sudden excessive increase in loading, deterioration failure is progressive and time-dependent. Consequently, the most vulnerable parts of the structure initially begin to exhibit minor failures such as cracking and spalling. As deterioration becomes severe, the spalled or corroded areas increase over time, eventually covering large parts of the structure. The foregoing scenario is suitably quantified by considering the probability of failure (P_f) at any given time. Hence, the likelihood that resistance is lower than loading at any given time, can be expressed as given in Equation (2.6). Obviously at time t = 0, the probability of corrosion failure is $P_f = 0$, but years later once/if corrosion has ultimately occurred throughout the structure, $P_f = 1.0$. The stochastic applicative method provides the methodology of calculating P_f values ranging from 0 to 1.0 at any given time during deterioration progression, as further discussed in the next Section 2.5.2.

$$P_f(t) = P\{R(t) - S(t)\} < 0 \tag{2.6}$$

The time-dependent changes in R(t) and S(t) values, are generally represented schematically as shown in Figure 2.3. Initially at t = 0, the two (2) distributions are far apart. But at time t = t_j, resistance decreases while loading increases causing the two (2) curves to approach each other. Subsequently at some point in time, the two (2) distributions overlap. The overlapping area represents failure probability, P_f which proportionally gives the likelihood of corrosion occurrence. The value, P_f increases with time until the maximum acceptable failure probability, P_{fmax} is reached at t = t_{sl}. Hence at ESL, Equation (2.6) can be written as Equation (2.7) (Sarja and Vesikari, 1996). As such, P_{fmax} represents the safety level which serves as the criterion for ESL.

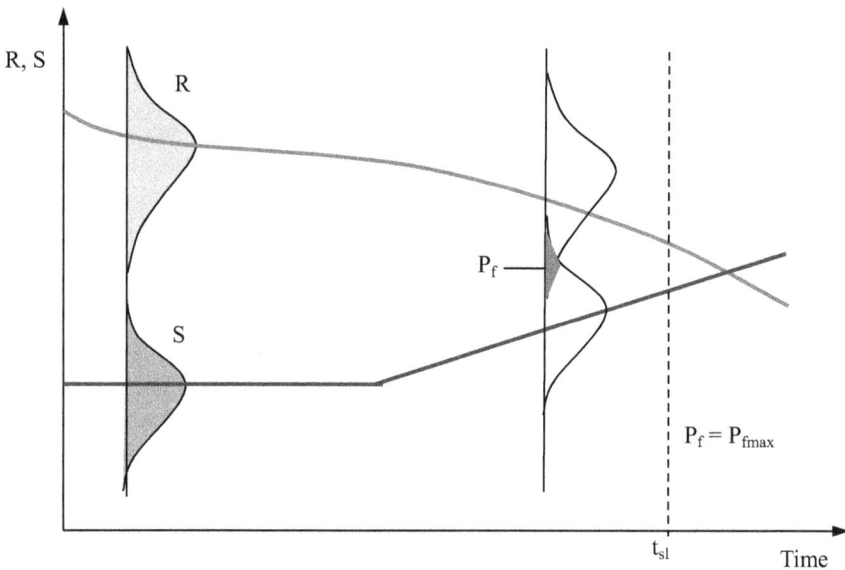

FIGURE 2.3 Increase in failure probability with time (Sarja and Vesikari, 1996).

$$P_f(t) = P\{R(t) - S(t)\} < P_{fmax} \tag{2.7}$$

2.5.2 RELIABILITY INDEX ANALYSIS

The stochastic applicative methodology is employed to determine failure probability, P_f at any given time of deterioration progression. The first step of the procedure is to quantify reliability of the load-resistance system by calculating a single numerical value, referred to as the *reliability index* (β). For time-dependent $R(t)$ and $S(t)$ quantities of normal distribution, the value of β is calculated using Equation (2.8).

$$\beta(t) = \frac{\mu(R,t) - \mu(S,t)}{\sqrt{\sigma^2(R,t) + \sigma^2(S,t)}} \tag{2.8}$$

where, $\mu()$ is the mean of and σ is standard deviation.

The value of P_f that corresponds to that of β can be read-off from standard normal distribution tables or calculated using the NORMDIST function of Excel spreadsheet. Equation (2.8) applies to scenarios wherein both R and S are time-dependent quantities. It is often the case that one (1) of the two (2) quantities R or S can be constant. When R is constant and S is time-dependent, Equation (2.8) can then be written as the simpler expression given in Equation (2.9).

$$\beta(t) = \frac{R - \mu(S,t)}{\sigma(S,t)} \tag{2.9}$$

2.5.3 APPLICATION OF RELIABILITY INDEX ANALYSIS TO CARBONATION

Carbonation attack process involves the penetration of CO_2 into the cover depth to reach the level of steel, and to consequently cause reinforcement corrosion. As such, failure occurs when the carbonation front reaches or exceeds the depth of cover location wherein steel reinforcement is embedded. Accordingly, loading S is carbonation depth while the resistance R is provided by cover depth. It is known that carbonation depth and cover depth are parameters that both exhibit the normal distribution curve. However, the average cover depth remains constant over time, while carbonation depth changes with time in accordance with the square root law model given in Equation (2.3f). Figure 2.4 illustrates the time-dependent progression of carbonation depth as a normal distribution curve, towards a constant cover depth (Sarja and Vesikari, 1996). The tail end of the normal distribution curve, would be the first part of the curve to exceed the cover depth. That portion of the curve's tail end which exceeds the cover depth, represents failure probability. Since cover depth is assumed to be constant, Equation (2.9) is the appropriate expression to use for calculation of failure probability. The reliability *index* analysis methodology is illustrated later using an example given in Chapter 6, Section 6.4.

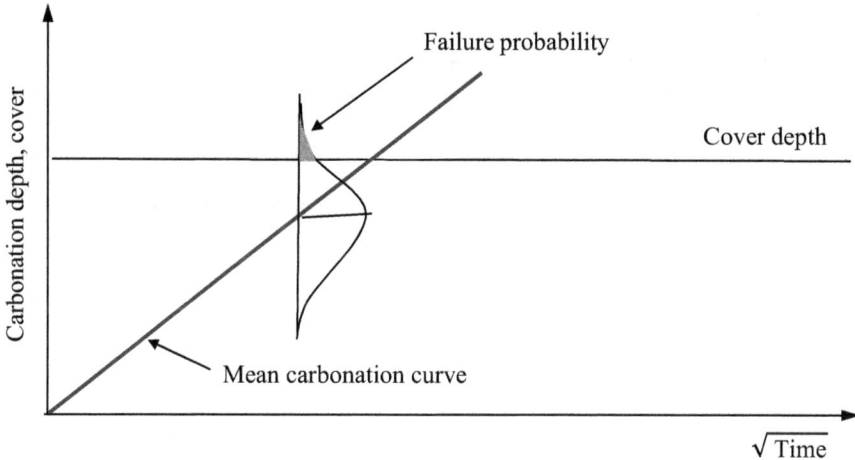

FIGURE 2.4 Carbonation progression into concrete cover leading to failure (Sarja and Vesikari, 1996).

2.5.4 MONTE CARLO SIMULATION OF CARBONATION

While both, reliability *index* analysis and the Monte Carlo simulation techniques are stochastic approaches, the latter is a computational method of determining failure probability by considering multiple possible outcomes governed by random variables. Both techniques require employment of a mathematical model relating the dependent variable which in this case is the carbonation progression, and the independent variables being the various factors influencing the mechanism. Each of the independent variables is assigned a suitable type of probability distribution that is representative of the parameter's natural variability tendency. The common types of such probability distributions are the normal, log-normal, uniform or triangular distributions, among others. Again, by definition, failure occurs once carbonation depth exceeds the cover depth, which can be written as the limit state Equation (2.10).

$$\text{Outcome} = c - d_{c,t}(X) < 0 \qquad (2.10)$$

where $d_{c,t}$ is carbonation depth, c is the cover depth and X represents the set of random independent variables or parameters. The outcome of Equation (2.10) is computed repeatedly to produce thousands of simulation runs (N). Failure probability is computed as given by Equation (2.11). Application of the Monte Carlo simulation technique is illustrated using an example given in Chapter 6, Section 6.5.

$$F(P) = \frac{\sum_1^N \left(c - d_{c,t}(X) < 0 \right)}{N} \qquad (2.11)$$

REFERENCES

Alexander M.G. (2018) Service life design and modelling of concrete structures – background, developments, and implementation, *Revista ALCONPAT*, 22, http://dx.doi.org/10.21041/ra.v8i3.325

Broomfield J.P. (2007) *Corrosion of steel in concrete: understanding, investigation and repair*, 2nd Ed, Taylor & Francis, London UK.

Cole W. and Kroone B (1960) Carbon dioxide on hydrated Portland cement, *Journal of American Concrete Institute*, September, 1275–1295.

CSIR (1999) *Final report on an investigation into the influence of fly ash on the durability of concrete*, by W.R. Barker, Division of Building and Construction Technology, CSIR, PO Box 395, 0001 Pretoria, South Africa, 18p. Also, Ash Resources Also, Ash Resources (Pty) Ltd, PO Box 3017, Randburg 2125.

Eder W.E. (2014) Chapter 10: an anthology of theories and models of design. In: *Engineering design: role of theory, models and methods*, Springer, London UK.

Ekolu S.O. (2018) Model for practical prediction of natural carbonation in reinforced concrete: part 1-formulation, *Cement and Concrete Composites*, 86, 40–56.

EN 14630 (2006) Determination of carbonation depth in hardened concrete by the phenolphthalein method. In: *Products and systems for the protection and repair of concrete structures – TEST methods*, European Committee for Standardization, CEN, Management Centre, Rue de Stassart, 36 B-1050 Brussels, Belgium, 5p.

Fulton's (2001) *Fulton's concrete technology*, Cement and Concrete Institute, Midrand, South Africa, 330p.

Hamada M. (1969) Neutralisation of concrete and corrosion of reinforcing steel. In: *5th international symposium on chem of cem*, III-3, 343–368.

Hunt C., Dantzler V., Tomes L.A. and Blaine R.L. (1958) Reaction of Portland cement with carbon dioxide, *Journal of Research of the National Bureau of Standards*, 60(5), 441–446.

Ikotun J.O. (2017) Effects of concrete quality and cover depth on carbonation-induced reinforcement corrosion and initiation of concrete cover cracking in reinforced concrete structures, *PhD Thesis*, School of Civil and Environmental Engineering, University of the Witwatersrand, 265p.

Klaus G. (1965) *Kybernetik in philosophischerSicht (Cybernetics in Philosophical View)*, 4th Ed, Dietz Verlag, Berlin.

Kropp J. (1995) Performance criteria for concrete durability. In: *Relations between transport characteristics and durability* (J. Kropp & H.K. Hilsdorf, Eds), E & FN Spon, London, UK.

Leber I. and Blakey F. (1956) Some effects of carbon dioxide on mortars and concrete, *Journal of American Concrete Institute*, September, 295–308.

Liang M.-T., Chang J.-J., Chang H.-T. and Yeh C.-J. (2012) Rust-expansion-crack service life prediction of existing reinforced concrete bridge/viaduct using time-dependent reliability analysis, *Journal of Marine Science and Technology*, 20(4), 397–409.

Neville A.M. (1996) *Properties of concrete*, 4th Ed (also 3rd Edition 1981), John Wiley and Sons Inc, New York.

Parrott L.J. (1987) A review of carbonation in reinforced concrete, *Cement and Concrete Association*, BRE, 42p.

RILEM CPC-18 (1988) *Measurement of hardened concrete carbonation depth*, TC56-MHM hydrocarbon materials, RILEM Recommendations, 3p.

Sagues A.A., Moreno E.I., Morris W. and Andrade C. (1997) *Carbonation in concrete and effect on steel corrosion*, Final Report, State Job No. 99700-3530-119, WPI 0510685, College of Engineering, University of South Florida, 4202 Fowler Avenue, Tampa, Florida 33620-5350, 299p.

Sarja A. and Vesikari E. (1996) *Durability design of concrete structures*, RILEM Report 14 (A. Sarja & E. Vesikari, Eds), E & FN Spon, UK, 93p.

Tuutti K. (1982) *Corrosion of steel in concrete*, Cement and Concrete Research Institute, Stockholm, 473p.

Verbeck G. (1958) *Carbonation of hydrated Portland cement*, ASTM Special Publication N205, 17–36.

3 The natural carbonation prediction (NCP) model and methodology

3.1 MODEL DEVELOPMENT

The NCP model was developed by the book author during his academic research at South African university institutions over a period of 12 years. But its formulation was fundamentally possible owing to past research and scientific advances reported in the literatures focussed on understanding concrete carbonation. The most important scientific advances in this regard have occurred progressively and cumulatively since the 1960s. Towards end of the 20th century, the following significant developments emerged: (i) scientific understanding of concrete carbonation and modelling using the Fick's laws governing the diffusion mechanism, (ii) generation of experimental and field data from real-life concrete structures. In the earlier years of attempting to understand concrete carbonation and factors that influence the mechanism, major focus was devoted to laboratory-based accelerated tests. Understanding of the various variables that influence carbonation was necessary, without which modelling would not be possible. It is now understood that the important variables affecting carbonation are concrete materials, concrete properties and environmental factors. Indeed, the following three (3) categories of experimental investigations have been responsible for carbonation data generation.

(a) Accelerated carbonation testing of pastes, mortars and concretes, to determine behaviour, response and factors of influence. In some cases, attempts were made to employ theoretical modelling concepts.
(b) Experimental studies on natural carbonation. As field observations became common experiences, with steel corrosion arising as a widespread durability problem of concrete, experimental set-ups intended to depict real-life behaviour became increasingly relevant. Such experimental data began to emerge in the 1980s and presently continue to be generated.
(c) Insitu carbonation data of real-life engineering structures. With more reinforced concrete structures undergoing corrosion-related repair works, data from tests conducted on real-life structures have been emerging from field-based investigations. Since such carbonation testing involves coring of the structure, availability of these data is particularly limited.

DOI: 10.1201/9781003645399-3

The NCP model employs the natural carbonation data of categories (b) and (c). No accelerated test data are involved on any aspects or components of the model. The modelling approach employed to formulate the NCP model was to express the natural phenomenon as a system of mathematical functions. Evidently, past advances on understanding concrete carbonation laid the foundation upon which the NCP model was developed. Among such research developments are the various experimental carbonation models in the literatures that informed the structuring of NCP model during its formulation.

3.2   VERSIONS OF THE MODEL

The first version of NCP model was published in Ekolu (2018) wherein its formulation is described comprising derivation, data fitting and optimization. This book does not delve into these aspects of the model's formulation. Rather, the original paper can be consulted for detailed information in this regard. Further to the original model, incremental research continued that led to the subsequent publications (Ekolu, 2020a, 2020b; Ekolu and Solomon, 2021). During this research process, however, the author also had the opportunity to better understand the model's limitations, from which arose the need to make improvements. To a great extent, the review comments received towards publication of manuscripts, were an excellent source of insights that led to improvements culminating into the comprehensive version of the NCP model presented in Section 3.4. The main improvements that were made are subsequently discussed.

3.3   UPDATES AND IMPROVEMENTS

An equation was developed which gives the relationship between the standard 28-day strength (f_{c28}) typically determined under laboratory curing conditions, and the *in situ* concrete strength (f_{cbn}) that is obtained from real-life structures exposed outdoors. The new equation enhances flexibility and versatility of the model, as it allows measurements taken at any given time period during the structure's service life, to be employed in design and/or analysis.

One of the major limitations of the original model published in Ekolu (2018), was lack of consideration to account for temperature effects of the natural climates. Incorporation of a temperature correction method would enhance versatility of the model, since concrete infrastructures are located in different climate regions worldwide. Also for service life design, the temperature submodel would provide corrective adjustments to dynamically account for time-dependent future CO_2-driven climate change impacts. Indeed, a temperature correction submodel was developed and embedded as part of the comprehensive NCP model (Section 3.4, Eq. 3.5). The development and description of the temperature submodel is given in Ekolu (2020a).

Other updates were also implemented to improve the model's flexibility and adaptability. The original model catered for only few cement types comprising CEM I and composites containing fly ash or slag. Later an equation (Section 3.4, Eq. 3.8b) for the scalar quantity, g, was developed for use to account for effects of different SCMs of varied proportions. Consequently, the comprehensive model caters for most varieties of cement composites, covering all the standard CEM I to V designations

(EN 197-1, 2000). Standard cement types typically contain different varieties and proportions of conventional SCMs blended with clinker cement secondarily to produce binary, ternary or quaternary composites.

3.3.1 RELATIONSHIP BETWEEN THE STANDARD 28-DAY STRENGTH AND *IN SITU* CONCRETE STRENGTH

The model's equations utilize, f_{c28} and/or f_{cbn}. By implication, f_{c28} and f_{cbn} should be useable interchangeably. Hitherto, however, no inter-relationship had been established between the two (2) parameters. Hence to enhance the model's versatility, it became essential to develop an equation relating f_{c28} and f_{cbn}.

Accordingly, data from the long-term 10-year carbonation investigation reported in CSIR (1999), were employed to determine the relationship between f_{c28} and f_{cbn}. Experimental details of the CSIR (1999) data are already given in Ekolu (2018) and only the relevant aspects are mentioned here for convenience. The data consisted of 25, 35 and 50 MPa concretes of 0.46 to 0.79 water/cementitious (w/cm) ratio. The cement types used in the mixtures comprised ordinary Portland cement (OPC or CEM I) and rapid hardening Portland cement (RHC) with or without 15%, 25%, 30% and 50% FA. Concrete cubes of 100 mm size were cast and subjected to a wide range of curing regimes comprising: (a) curing in a fog room for 28 days, (b) curing at 80% RH/50°C for 24 hours followed by 50% RH/10°C curing for 27 days, (c) curing at 50% RH/23°C for 28 days, (d) steam curing for four (4) hours at 60°C followed by 50% RH/23°C curing for 27 days, (e) oven-dry curing for 24 hours at 40°C followed by 50% RH/23°C curing for 27 days. After 28 days of curing as per the foregoing regimes (a) to (e), the cube samples were exposed outdoors at the roof of a storeyed building under Pretoria weather in South Africa. The outdoor-exposed cube samples were then tested for natural carbonation and for compressive strength at the ages of 3.5, 6 and 10 years.

Since the model was focused on applicability to real-life structures, the data selected for use in the study, were those that depicted engineering practice. As such, regime (a) involving 28 days of continuous fog room curing, was excluded, as this procedure does not represent construction site practice. Since construction specifications, standards and guidelines (BS 8110, 1997; SANS 10100, 2000) typically recommend site curing durations of three (3) to five (5) days (Ekolu, 2016), regimes (b) to (e) were selected for employment in the study, considering that these treatments somehow depict the partial saturation curing condition representative of construction practice. Moreover, the curing regimes (b) to (e) are also generally consistent with concrete precasting and heat treatment practice.

Literatures provide different types of recognized models for prediction of long-term concrete strength growth with age. Shariq et al. (2010) reported that the strength prediction models given in CEB-FIP (1990), GL-2000 (2001) and ACI 209 (1997), all showed comparative results. However, these models apply to strength growth under the saturated moist-curing condition, but not under the outdoor exposure condition at which natural carbonation occurs in real-life structures. Under the ideal moist-curing condition comprising immersion storage of samples in water, full saturation of concrete occurs, which fully promotes cement hydration to give maximum strength gain. However, concrete in structures is typically partially saturated under the seasonal fluctuating outdoor weather conditions of temperature, RH, precipitation, sunlight and radiations. Also, natural carbonation of concrete during outdoor exposure, leads to strength enhancement. In contrast, no carbonation occurs during saturated moist-curing of concrete. Clearly, the real-life outdoor environmental conditions to which structures are exposed, are vastly different, and expectedly affect concrete strength in a manner different from that of the controlled laboratory conditions used to determine the standard 28-day strength. The foregoing considerations further underscored the need to develop the relationship between f_{c28} and f_{cbn}.

Using the CSIR (1999) data already described earlier, the new equation given in Figure 3.1 was developed for prediction of f_{cbn} from strength results of f_{c28} or vice

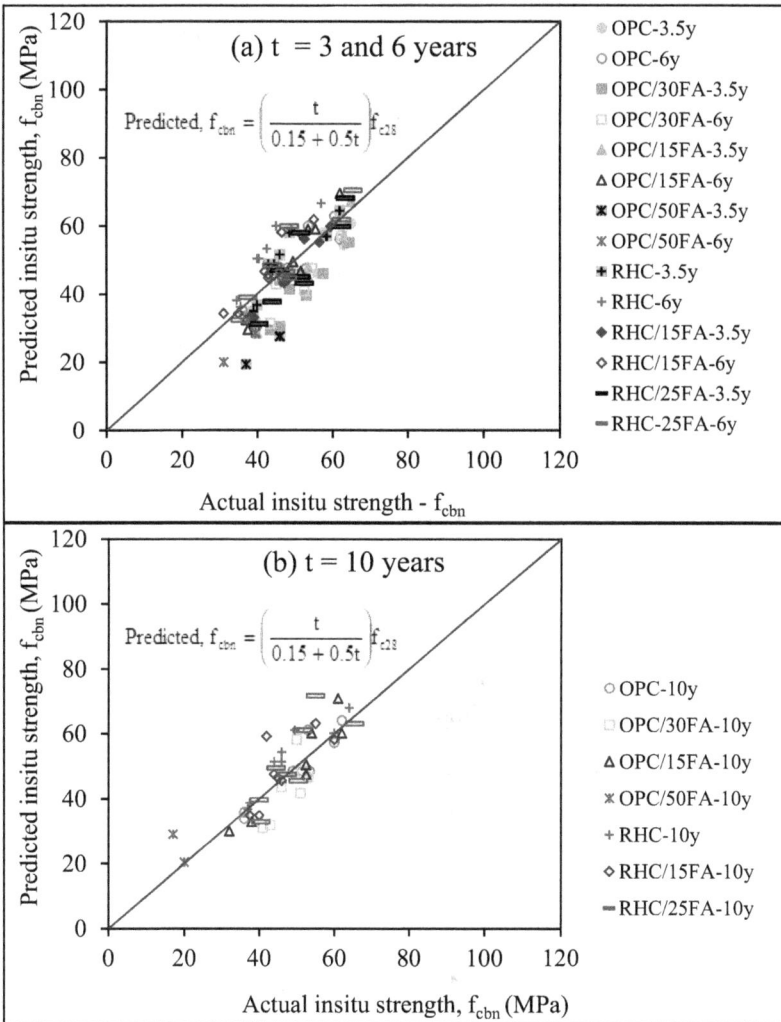

FIGURE 3.1 Comparison of predicted *in situ* strength versus actual *in situ* strength results for (a) ages of 3.5 and 6 years, (b) age of 10 years: OPC/30FA-6y is the six (6) year compressive strength of concrete made with ordinary Portland cement (OPC) containing 30% fly ash (FA), RHC/15FA-10y is the ten (10) year compressive strength of concrete made with rapid hardening Portland cement (RHC) containing 15% FA, etc.

versa. This relationship equation applies to concrete mixtures containing $\leq 30\%$ FA or common SCMs, along with those containing $\leq 50\%$ slag. A plot of predicted versus actual values of f_{cbn} as seen in Figure 3.1, gives data points which closely fall along the line of equality, thereby depicting a strong agreement between the two (2) sets of results. The new equation allows, f_{c28} and f_{cbn} to be employed interchangeably in the model's equations given later in Section 3.4.

The development of f_{c28} to f_{cbn} conversion equation significantly enhanced the versatility of NCP model. Indeed, the model's Equations (3.3) to (3.6) given in Section 3.4 are based on f_{c28}. But for existing structures wherein f_{cbn} is the *in situ* strength that can be measured, it would be difficult or impossible to effectively apply the model without employing the f_{cbn} to f_{28} conversion equation.

3.3.2 RELATIONSHIP BETWEEN 28-DAY MEAN STRENGTH
AND CONCRETE STRENGTH GRADE

In design of concrete structures, the CEB-FIB model code recommends use of the relationship $f_c = f_{ck}+8$ (CEB-FIP, 2010; EN 1992-1-1, 2005). This equation can be particularly useful at design stage of new structures wherein strength grade is typically used, rather than employment of mean or target strength values. The CSIR (1999) data described in Section 3.3.1 were employed to evaluate veracity of this standard equation for potential use in service life modelling, as recommended by Sarja and Vesikari (1996).

The 25, 35 and 50 MPa concretes gave mean strengths of 25, 35 and 51 MPa under the moist-curing regime, confirming that these mixtures were well-designed to achieve their intended or target strength values. Moreover, the results included mixtures containing 30% to 50% FA. The respective strength grades of moist-cured concretes were determined to be 16, 23.5 and 45 MPa. But concretes that had been subjected to curing regimes (b) to (e) gave diminished 28-day strength values, along with lower strength grades of 12, 18 and 23 MPa for the 23, 35 and 50 MPa concretes, respectively.

The values for f_{c28} and $f_{ck}+8$ were calculated for the reference moist-curing regime (a) and for all the other curing conditions (b) to (e), then used to evaluate the relationship between the two (2) parameters. A total of 94 data points were employed in the evaluation. Figure 3.2 shows the graph of f_{c28} versus $f_{ck}+8$. It can be seen in Figure 3.2 that the data points lie along the line of equality, which indeed shows that the two (2) strength parameters can be used interchangeably without loss of accuracy (Sarja and Vesikari, 1996; CEB-FIP, 2010).

3.3.3 ACCOUNTING FOR TEMPERATURE EFFECT

As mentioned at the beginning of Section 3.3, a temperature correction submodel was developed to account for the effect of climate temperature or related temperature changes, on concrete carbonation. Since it is difficult or impossible to set-up a natural carbonation experiment at which a constant temperature is maintained outdoors, an accelerated carbonation study simulating natural carbonation is the best option.

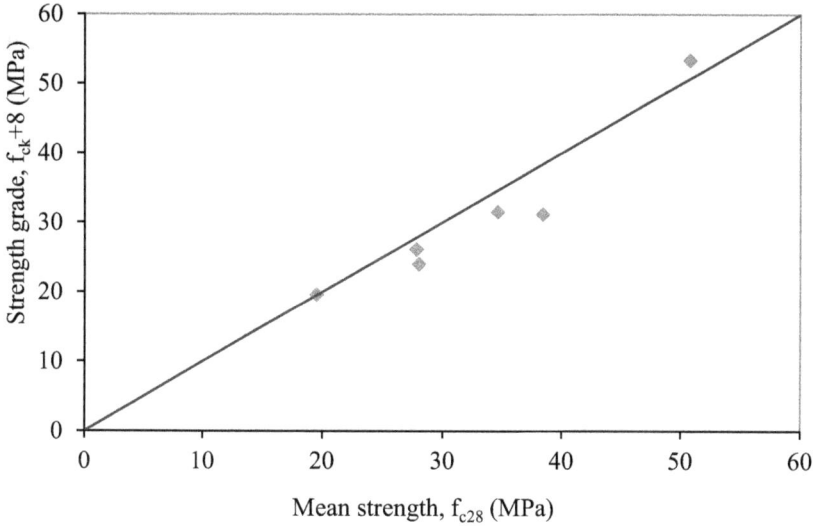

FIGURE 3.2 Relationship between mean strength (f_{c28}) and characteristic strength grade parameter ($f_{ck}+8$).

Accordingly, the accelerated carbonation data reported by Peng et al. (2018) were employed to develop the model, as described in Ekolu (2020a). The data comprised results of 25, 30 and 35 MPa concretes subjected to accelerated carbonation at varied temperatures of 20 to 40°C, while exposed to 20% CO_2 concentration. In the analysis, results obtained at 20% CO_2 concentration were converted to their equivalent under natural carbonation, based on experimental correlations derived from the literatures (Ji et al., 2010; Neves et al., 2013). Subsequently, the Arrhenius-type temperature correction equation given in Figure 3.3 was developed. An excellent correlation between the calculated carbonation and the actual natural carbonation, can be seen in Figure 3.3.

3.4   DESCRIPTION OF THE NCP MODEL

Ekolu (2018) gives description of the original model, its derivation and optimization. Subsequent incremental research led to new developments and model improvements (Section 3.3). Accordingly, new equations were incorporated into the original model, to formulate the comprehensive version comprising a system of Equations (3.1) to (3.9). Updates that were incorporated into the model, are subsequently discussed.

The original version of the model (Ekolu, 2018) was applicable only to concrete structures located in the tropical climate regions. This major limitation could not allow the original model to be employed across worldwide climates of the vastly different environmental conditions, ranging from the temperate extreme cold winter climate, to the tropical hot and humid summer weather. To resolve this limitation, a temperature correction submodel given as Equation (3.5), was incorporated (Section 3.3.3).

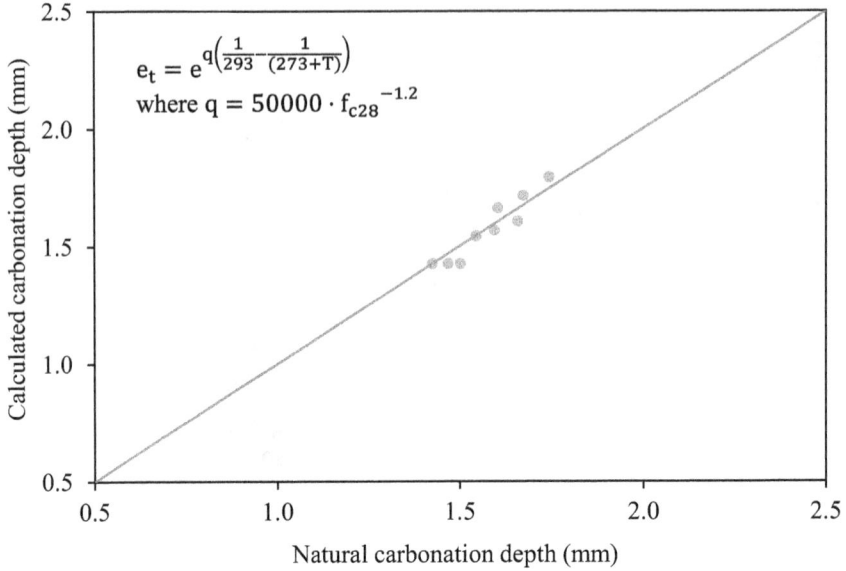

$$e_t = e^{q\left(\frac{1}{293} - \frac{1}{(273+T)}\right)}$$

where $q = 50000 \cdot f_{c28}^{-1.2}$

FIGURE 3.3 Correction for temperature effect: e_t is the temperature correction factor, T is annual ambient temperature in °C at the site location considered, f_{c28} is 28-day concrete strength.

It is also common knowledge that a majority of the cements used in modern engineering construction works, contain SCMs in accordance with standard specifications (EN 197-1, 2000; ASTM C1157, 2017). The original model (Ekolu, 2018) was applicable for only few cement types comprising CEM I and some composites containing fly ash and blast-furnace slag. Improvement of the original model led to the comprehensive model which has enhanced features including provisions that cater for most varieties of composites covering all the standard cement designations CEM I to V of EN 197-1 (2000). The vast range of cement types specifically catered for in the updated model are CEM I and CEM II/A,B containing the various common SCMs (slag, silica fume, pozzolana, fly ash, burnt shale, limestone and their composites), CEM III/A, CEM IV/A and CEM V/A. In the model's Equation (3.8a, b), the various cement types are grouped into three (3) categories comprising those composites containing ≤ 20% SCM(s) (CEM I, CEM II/A), containing ≤ 30% SCMs (CEM II/B, CEM IV/A and CEM V/A), or containing ≤ 50% slag (CEM III/A).

Also, the tabulated cement factors (*cem, g*) of Equation (3.8a) are complemented with the more general Equation (3.8b) that can be used as an alternative to the values given in the table. The cement types that are not presently catered for in the NCP model, are those containing unusually high-volume proportions comprising > 50% SCMs which are CEM III/B,C; CEM IV/B and CEM V/B varieties. It may be highlighted that these particular cement types containing such high-volume proportions of SCMs, are generally not specified for structural use in reinforced concrete, since such composites typically exhibit weak performance characteristics including high creep and shrinkage, high carbonation and low strength.

The model's Equations (3.3) to (3.6) are based on f_{c28} while Equation (3.7) provides the option of utilizing f_{cbn}, which is the *in situ* strength determined from existing structure(s). Equations (3.3) to (3.6) that employ f_{c28}, only apply to new structures or existing structures that have historical data of 28-day strength results. However, most existing or old structures typically may not have historical records of f_{c28} results, which would make the original model impossible to apply. This issue was one of the important limitations of the original model (Ekolu, 2018). As a provision to adapt the model's applicability universally to both new and existing or old concrete structures, the relationship between f_{c28} and f_{cbn} was developed (Section 3.3.1) and added as Equation (3.9c) of the updated comprehensive model. The new equation allows conversion of f_{c28} values to equivalent f_{cbn} results, and vice versa. Consequently, the updated model can be employed for design and analysis of new and existing concrete structures alike, provided any of the strength values comprising f_{c28} or f_{cbn} is known or can be determined. Equation (3.9a-c) gives a set of three (3) necessary conditions that must be consistently observed when employing the NCP model.

Carbonation depth function, $d_{(c,t)} = e_h \cdot e_s \cdot c_c \cdot e_t \cdot cem \left(F_{c(t)} \right)^g \sqrt{t}$ (3.1)

where e_h, e_s, e_c and e_t are environmental correction factors for RH, sheltering, CO_2 concentration and temperature, respectively. $F_{c(t)}$ is the function for strength growth with time (t), which in turn is converted into carbonation progression using the scalar quantity, *cem*, coupled with exponent, *g*, both factors being dependent on the type of cement.

Environmental correction factor for relative humidity (RH)

$$e_h = 16 \left(\frac{RH - 35}{100} \right) \left(1 - \frac{RH}{100} \right)^{1.5} \text{ for } 50\% \leq RH \leq 80\%$$ (3.2)

Environmental correction factor for sheltering

$$e_s = \begin{cases} 1.0 \text{ for sheltered outdoor exposure} \\ f_{c28}^{-0.2} \text{ for unsheltered outdoor exposure; } f_{c28} \text{ is } 28-\text{day strength} \end{cases}$$ (3.3)

Environmental correction factor for CO_2 concentration

$$e_c = \begin{cases} f_{c28}^r \text{ for } 20 < f_{c28} < 60 \text{MPa } \alpha \\ 1.0 \text{ for } f_{c28} \geq 60 \text{MPa} \end{cases}$$ (3.4)

where α, r are given in the following for natural carbonation under varied CO_2 concentrations

28-day strength (MPa)	Correction factor		CO_2 concentration level (ppm)				
			200	300	500	1000	2000
$20 < f_{c28} < 60$	$e_c = \alpha f_{c28}^r$	α	1.4	1.0	2.5	4.5	14
		r	$-1/4$	0	$-1/4$	$-2/5$	$-2/3$
$f_{c28} \geq 60$	$e_c = 1.0$						

Environmental correction factor for temperature

$$e_t = e^{q\left(\frac{1}{293} - \frac{1}{(273+T)}\right)}, \text{ where } q = 50000 \cdot f_{c28}^{-1.2} \tag{3.5}$$

The model's reference annual temperature is 20°C. T is the annual ambient temperature in °C at the site location considered, f_{c28} is the 28-day concrete strength.

Time-dependent strength growth function ($F_{c(t)}$)

$$F_{c(t)} = \frac{t}{a + bt} \cdot f_c$$

where $f_c = f_{c28}$ or f_{cbn} \hfill (3.6a)

   (a) Using 28-day strength (f_{c28})
       (i) Short-term ages, $t < 6$ years

$$a = 0.35, b = 0.6 - t^{0.5} / 50 \tag{3.6b}$$

       (ii) Long-term ages, $t \geq 6$ years

$$a = 0.15t, b = 0.5 - t^{0.5} / 50 \tag{3.6c}$$

   (b) Using long-term *in situ* strength (f_{cbn})

       (i) Short-term ages, $t < 15$ years

$$a = 0.35, b = 1.15 - t^{0.6} / 50 \tag{3.7a}$$

       (ii) Long-term ages, $t \geq 15$ years

$$a = 0.15t, b = 0.95 - t^{0..6} / 50 \tag{3.7b}$$

Cement factors (*cem, g*) for carbonation conductance \hfill (3.8a)

SCMs*	Cement types	Scalar, *cem*	Conductance factor, *g*
≤ 20% common SCMs	CEM I, CEM II/A varieties	1000	−1.5
≤ 30% common SCMs	CEM II/B varieties, CEM IV/A, CEM V/A	1000	−1.4
≤ 50% slag	CEM III/A	1000	−1.4

Note: *SCMs are common or conventional supplementary cementitious materials: blast-furnace slag, silica fume, pozzolana, fly ash, burnt shale, limestone, composites of SCMs. If the proportion of the SCM (%SCM) in Portland cement is known, Equation (3.8b) may be used to determine, *g*.

$$cem = 1000, g = \frac{\%SCM}{500} - 1.5, \text{ where } \%SCM \text{ is } \leq 30\% SCMs \text{ or } \leq 50\% slag \qquad (3.8b)$$

Conditions for application of the model

- Cube strength (f_c) is related to core or cylinder strength (f_{cyl}) as per the conversion, $f_c = 1.25 f_{cyl}$. (3.9a)
- Cube strength values used in the equations must be ≥ 20 MPa. (3.9b)
- The 28-day cube strength (f_{c28}) is related to *in situ* strength (f_{cbn}) in accordance with the equation below. In cases such as existing structures, wherein f_{cbn} may be the only known strength parameter, this conversion equation can be employed to enable implementation of the model's Equations (3.3) to (3.6).

$$f_{cbn} = \left(\frac{t}{0.15 + 0.5t} \right) f_{c28} \qquad (3.9c)$$

3.5   APPLICATIONS AND LIMITATIONS

3.5.1   RANGE OF APPLICATIONS

Later in Chapters 4 and 5, it is shown that the NCP model has been subjected to extensive and severe testing to verify its performance, accuracy, robustness, veracity and applicability worldwide. Refinements and improvements were done upon the model. By evaluating and testing the model using a wide range of data sets from worldwide sources, its scope of applicability was established along with its limitations. Accordingly, the NCP model is applicable to:

1. Natural carbonation. Accelerated carbonation is not relevant to the model.
2. Structural normal concretes of cube strengths ≥ 20 MPa.
3. Outdoor exposure environments. The present version of the model does not apply to indoor exposure conditions.
4. Global climates comprising the temperate, tropical and subtropical regions worldwide. It may be noted that these climates are the geographical regions

which naturally have habitable environmental conditions, and wherein major world cities and urban infrastructures are located.

5. New and existing or old concrete structures of all ages.
6. Concretes made with normal aggregate types. The present NCP model version does not account for use of special aggregate types such as lightweight, heavyweight or recycled aggregate.
7. Uncracked concrete. Cracks provide a direct CO_2 entry path, rather than gas diffusion into concrete.
8. Concretes free of surface protection treatments such as paints and coatings. Future research would be needed to develop the adjustment factors that can be used to account for effects of surface treatments.

3.5.2 APPLICATION TO DESIGN AND ANALYSIS OF NEW CONCRETE STRUCTURES

The NCP model can be employed for durability design to determine suitable specification requirements such as concrete strength, the cementitious material system and cover depth, among others. The model can also be used to conduct service life design and analysis, as demonstrated later in Chapter 6. Moreover, the model allows employment of two (2) different strength parameters comprising the strength grade of concrete or mean 28-day strength. Often during design of new concrete structures, only the concrete strength grade is known. The NCP model can still be employed using the strength grade parameter for durability design and for service life analysis. However, if concrete mixtures and strengths are known at design stage, it would be preferable to use the 28-day cube strength values determined from laboratory testing.

3.5.3 APPLICATION TO ANALYSIS OF EXISTING CONCRETE STRUCTURES

For existing concrete structures, it would be irrelevant to carry out durability design as the key parameters including strength, the cementitious material system and cover depth, would have been prescribed during construction. Instead, the NCP model can be employed for analysis of carbonation progression and for service life analysis of existing structures. The *in situ* concrete strength, f_{cbn} is particularly employed for analysis of existing concrete structures, especially when no records of the 28-day concrete strength may be available, as the case is with most old structures. The model's Equation (3.9c) allows conversion of f_{cbn} to f_{c28} and vice versa, making the model's employment more versatile. If the structure's age is known, the residual service life of the structure can be determined using the model.

3.6 PROCEDURE AND EXAMPLE OF THE MODEL-BASED CALCULATIONS

This section provides an example of calculations that illustrate the methodology of correctly implementing the NCP model. Users and readers can use this example as an illustration of the step-by-step procedure given in the schematic flow chart shown in Figure 3.4. Table 3.1 presents the data used in the example. Manual calculations

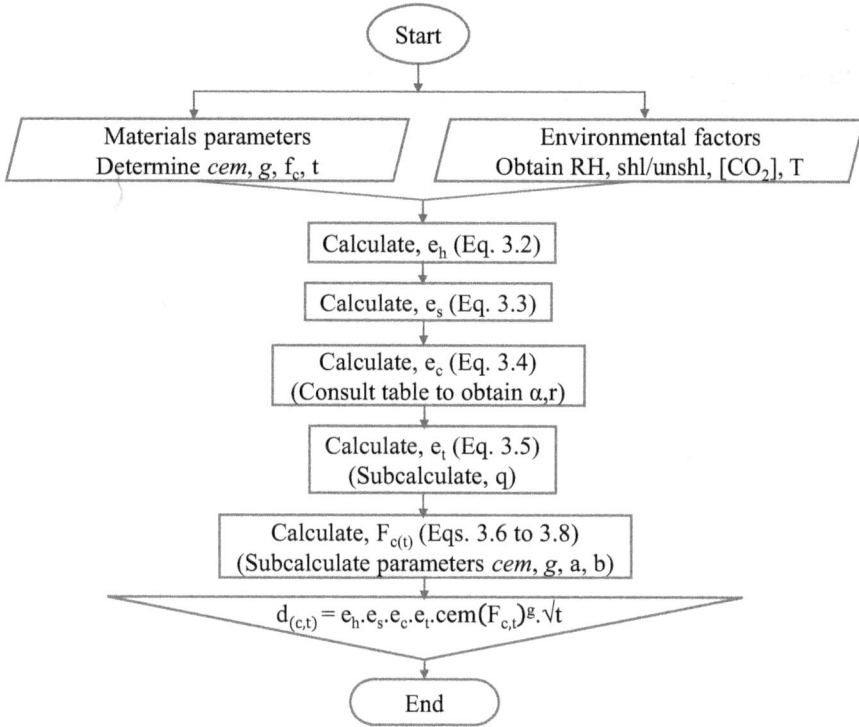

FIGURE 3.4 Schematic flow chart giving the step-by-step calculation procedure of NCP model.

have been done at an accuracy level of three (3) decimals. In the example, carbonation depth is calculated using the data given.

Carbonation depth function, $d_{(c,t)} = e_h \cdot e_s \cdot e_c \cdot e_t \cdot cem\left(F_{c,t}\right)^g \cdot \sqrt{t}$ (Eq. 3.1)

$$e_h = 16\left(\frac{RH-35}{100}\right)\left(1-\frac{RH}{100}\right)^{1.5} = 16\left(\frac{71.5-35}{100}\right)\left(1-\frac{71.5}{100}\right)^{1.5} = 0.889 \qquad \text{(Eq. 3.2)}$$

$$e_s = \begin{cases} 1.0 \text{ for sheltered outdoor exposure} \\ f_c^{\,0.2} \text{ for unsheltered outdoor exposure; } f_c \text{ is 28 - day cube strength} \end{cases} = 1.000 \qquad \text{(Eq. 3.3)}$$

$$e_c = \begin{cases} \alpha f_c^{\,r} \text{ for } 20 < f_c < 60MPa \\ 1.0 \text{ for } f_c \geq 60MPa \end{cases} = 1.75(53.7)^{\,0.125} = 1.064 \qquad \text{(Eq. 3.4)}$$

(By interpolating values in the table of Eq. (3.4) for $[CO_2]$ of 400 ppm, $\alpha = 1.75$, $r = -0.125$)

TABLE 3.1
Natural carbonation data used in the example.

Concrete properties	1. Cement type CEM II/A-LL containing 16% LL
	2. 28-day cube strength of 53.7 MPa
	3. Age of testing of two (2) years
Environmental conditions	4. *Sheltered* outdoor exposure
	5. CO_2 concentration of 400 ppm
	6. Average annual ambient relative humidity of 71.5%
	7. Average annual ambient temperature of 13.8°C

$$e_t = e^{q\left(\frac{1}{(273+T_{ref})} - \frac{1}{273+T_i}\right)}, \text{where, } q = 50000 \times f_{c28}^{-1.2} = 50000.53.7^{-1.2} = 419.760 \quad (Eq. 3.5)$$

$$\text{Therefore, } e_t = e^{q\left(\frac{1}{(293)} - \frac{1}{273+T_i}\right)} = e^{419.760\left(\frac{1}{(293)} - \frac{1}{273+13.8}\right)} = 0.970$$

$$F_{c(t)} = \frac{t}{a+bt} \cdot f_c, \text{where } f_c = f_{c28} \text{ or } f_{cbn} \quad (Eq. 3.6a)$$

$$a = 0.35, b = 0.6 - t^{0.5}/50 = 0.6 - 2^{0.5}/50 = 0.572 \quad (Eq. 3.6b)$$

$$F_{c(t)} = \frac{t}{a+bt} \cdot f_c = \frac{2}{0.35 + 0.572*2} \cdot 53.7 = 71.888$$

$$cem = 1000, g = \frac{\%SCM}{500} - 1.5 = \frac{16}{500} - 1.5 = -1.468 \quad (Eq. 3.8a, b)$$

From Eq. (3.1),

$$\text{Carbonation depth, } d_c = (0.889) \cdot (1.000) \cdot (1.064) \cdot 1000(71.888)^{-1.468} \cdot \sqrt{2} = 2.516 \text{ mm}$$

REFERENCES

ACI 209 (1997) *American Concrete Institute (ACI) Committee 209, Subcommittee II. Prediction of creep, shrinkage and temperature effects in concrete structures*, Report ACI 209 R92, (re-approved 1997).

ASTM C1157 (2017) *Standard performance specification for hydraulic cement*, ASTM International, West Conshohocken, PA.

BS 8110 (1997) *Structural use of concrete, part 1. Code of practice for design and construction, London*, British Standards Institution, London, UK.

CEB-FIP (1990) *Model code: design code*, Thomas Telford, London.

CEB-FIP (2010) *Model code 2010, first complete draft – volume 2, fib bulletin 56*, International Federation for Structural Concrete (fib), Case Postale 88, CH-1015, Lausanne, Switzerland.

CSIR (1999) *Final report on an investigation into the influence of fly ash on the durability of concrete* (W.R. Barker, Ed.), Division of Building and Construction Technology, CSIR, Pretoria, South Africa, 18p. Also, Ash Resources (Pty) Ltd, PO Box 3017, Randburg 2125.

Ekolu S.O. (2016) A review on effects of curing, sheltering, and CO_2 concentration upon natural carbonation of concrete, *Construction and Building Materials*, 127, 306–320.

Ekolu S.O. (2018) Model for practical prediction of natural carbonation in reinforced concrete: part 1-formulation, *Cement and Concrete Composites*, 86, 40–56.

Ekolu S.O. (2020a) Implications of global CO_2 emissions on natural carbonation and service lifespan of concrete infrastructures – reliability analysis, *Cement and Concrete Composites*, 114, 103744, http://doi.org/10.1016/j.cemconcomp.2020.103744

Ekolu S.O. (2020b) Model for natural carbonation prediction (NCP): practical application worldwide to real life functioning concrete structures, *Engineering Structures*, 224, 111126. http://doi.org/10.1016/j.engstruct.2020.111126

Ekolu S.O. and Solomon F. (2021) A case study on practical prediction of natural carbonation for concretes containing supplementary cementitious materials, *KSCE Journal of Civil Engineering*, 26(3), 1163–1176. https://doi.org/10.1007/s12205-021-1770-6

EN 197-1 (2000) *Cement – part 1: composition, specifications and conformity criteria for common cements*, European Committee for Standardization, CEN, 29p.

EN 1992-1-1(2005) *Eurocode 2: design of concrete structures – part 1-1: general rules and rules for buildings*, European Committee for Standardization, CEN, Management Centre, Rue de Stassart, 36 B-1050, Brussels, 225p.

GL-2000 (2001) Design provisions for drying shrinkage and creep of normal strength concrete, by Gardner, N. and Lockman, *ACI Materials Journal*, 98, 159–167.

Ji Y., Yuan Y., Shen J., Ma Y. and Lai S. (2010) Comparison of concrete carbonation process under natural condition and high CO_2 concentration environments, *Journal of Wuhan University of Technology – Materials Science Edition*, 25(3), 515–522.

Neves R., Branco F. and de Brito J. (2013) Field assessment of the relationship between natural and accelerated concrete carbonation resistance, *Cement and Concrete Composites*, 41, 9–15.

Peng J., Tang H., Zhang J. and Cai S.C.S. (2018) Numerical simulation on carbonation depth of concrete structures considering time- and temperature-dependent carbonation process, *Advances in Materials Science and Engineering*, Hindawi, Article ID 2326017, 16p., https://doi.org/10.1155/2018/2326017

SANS 10100–1&2 (2000) *Code of practice for the structural use of concrete, part 1: design; part 2: materials and execution of work*, South African Bureau of Standards Ltd, Pretoria.

Sarja A. and Vesikari E. (1996) *Durability design of concrete structures*, RILEM Report 14 (A. Sarja & E. Vesikari, Eds), E & FN Spon, UK, 165p.

Shariq M., Prasad J. and Masood A. (2010) Effect of GGBFS on time dependent compressive strength of concrete, *Construction and Building Materials*, 24(8), 1469–1478.

4 Experimental justification of the NCP model

4.1 CARBONATION-INFLUENCING FACTORS

A plethora of carbonation models have been proposed since the 1960s, most of them being experimental equations. While knowledge accumulation has advanced the various approaches, the carbonation modelling problem has not been resolved since 1960's, owing to its complexity. Past literatures have shown that a wide range of interacting factors influence carbonation. For purposes of carbonation modelling, it is necessary to express these factors as mathematical functions or empirical formulae (Ekolu, 2018).

It has been well-documented over the past decades that the carbonation-influencing factors fall into two (2) categories comprising the materials composition and environmental parameters. The compositional factors that influence carbonation are themselves wide ranging, comprising both chemical and material physical properties of the cementitious system. The main chemical compositional factors that are known to influence carbonation comprise: clinker content and products of cement hydration including CaO content, concentration of calcium hydroxide [CH], concentration of calcium silicate hydrate [CSH] and alkaline content of concrete (Parrot, 1987; CEB, 1997, Papadakis et al., 1992). The unhydrated cement phases comprising C_2S, C_3S, C_3A, C_4AF and degree of hydration, also have implications on carbonation (Ekolu, 2018). The concrete mixture properties affecting carbonation are perhaps even more numerous, with the most commonly identified factors ranging from mix design parameters to physical properties comprising water/cementitious ratio (w/cm), cement type, cement content, air content, curing, age, diffusion or permeability, compressive strength, porosity, among others. It also follows that serviceability conditions that influence material behaviour, also impact carbonation. Under this sub-category belong cracks, surface treatments such as paints and coatings of various kinds, along with synergistic presence of other degradation mechanisms including abrasion, chloride attack, freeze-thaw damage, alkali-silica reaction, sulphate attack, among others. Serviceability factors are, however, not core considerations in modelling but can be considered secondarily for specific applications of interest. It is also prudent and practical to consider carbonation modelling in isolation from other damage mechanisms.

The other category of crucial factors that majorly influence concrete carbonation is that of environmental parameters. These factors too are wide ranging of which the major ones known to be indispensable comprise: sheltering, rain or precipitation, RH, temperature and atmospheric CO_2 concentration. Other environmental factors including sun radiation and related weathering effects, etc., may be less significant.

DOI: 10.1201/9781003645399-4

TABLE 4.1
Categories of carbonation-influencing factors.

Category	Sub-category	Parameters and properties
Materials	Compositional factors	CaO content, [CH], [CSH], alkaline content.
	Physical factors	Water/cementitious ratio (w/cm), cement type, cement content, air content, curing, age, diffusion or permeability, compressive strength, porosity.
	Serviceability factors	Cracks, surface treatments including paints and coatings, other degradation processes.
Environmental exposure	Environmental factors	Sheltering, rain or precipitation, relative humidity (RH), CO_2 concentration, temperature.

Meanwhile, concrete can be as different as each batch of the mixture. That is, any of the factors associated with processing of concrete including mixing, casting, compaction, placement, finishing and curing, can easily affect mixture properties differently.

Table 4.1 gives a summary of the different categories of factors known to significantly influence concrete carbonation. Given such a wide range of factors, it is impossible to utilize all of them in the same model. This raises another complication of deciding which factors would be suitable for employment in modelling. Moreover, the process of identifying the suitable factors for modelling is not random, but has to be based on scientific observations and engineering behaviour of the cementitious system. Accordingly, research studies must strictly depend on scientific data and trends to guide the modelling process.

4.2 DATA AND MODELLING

Most standardized code-type engineering methods that are conventionally employed in practice, are typically analytical or empirical models. Owing to the practical nature of engineering, empirical models are based on or aim to depict realistic observations. Since engineering data patterns are typically governed by natural physical laws, it follows that the principles that underlie empirical models are based on theoretical laws of physics and mathematical functions. In several cases, such natural laws may or may not have been discovered theoretically through physics and mathematics, but data can unmistakably display the patterns or trends depicting real-life behaviour. Theoretically, physical processes of engineering can be expressed in form of mathematical equation(s). Hence by employing data of a given physical process and expressing it in form of mathematical function(s), an empirical model can be developed. But firstly, the appropriate mathematical function or general form of the equation must be identified, following which it can be adjusted to fit the data, for example, by employing statistical techniques such as curve-fitting and regression.

Modern advances in computational intelligence (CI) have led to emergence of an approach, referred to as data-driven modelling (DDM) (Solomatine et al., 2009). Physics laws and mathematical equations are not employed in DDM. The established CI methods include neural networks (NN), fuzzy systems and machine learning. NNs for example, comprise a network of interconnections between inputs, internal variables and outputs. A training data set is used to minimize or effectively eliminate large errors among values of interconnected variables between inputs and outputs. Once sufficiently trained, the NN system is then tested using another independent data set, to determine whether its outputs realistically predict the actual results. When properly configured, such CI systems can be effective. However, the absence of scientific fundamentals in CI methodology, is a disadvantage that undermines its implementation as a standalone technique of engineering.

As evident in the foregoing discussion, data are foundational to empirical or CI-based modelling. Accordingly, engineering models can only be as realistic as the quality of data employed in their development. Experimental models that are developed based on accelerated testing, may not represent the behaviour of real-life structures unless a practical relationship is established between the accelerated test values and natural results. Moreover, no single data set of carbonation experiments can be expected to sufficiently include comprehensively all the factors listed in Table 4.1. Hence it is often the case that models developed based on one (1) set of data generated under particular conditions, fail when applied to another data set obtained under a different set of conditions. Consequently, numerous experimental models fail to universally capture the complexity that underlies carbonation progression in concrete.

4.3 EXPERIMENTAL DATA OF NATURAL CARBONATION

Carbonation research studies are typically based on laboratory set-ups or field experiments. This book is only concerned with natural carbonation modelling, which is based on data from field-based experiments or measurements. Such experiments typically involve casting of concrete samples such as cubes, prisms or structural elements comprising beams, slabs or columns, etc., followed by their exposure outdoors to undergo natural carbonation. These field-based experiments for natural carbonation, are long-term studies typically lasting several years. Consequently, there are relatively few such studies, hence the scarcity of natural carbonation data. CSIR (1999) was perhaps the earliest and one of the longest natural carbonation field-based outdoor experimental studies, having lasted ten (10) years.

Considering that carbonation is a mechanism that occurs naturally worldwide, it is essential for practical engineering models to be universally applicable for employment globally. Across the world, industry utilizes different types of concretes, different cementitious systems, admixtures and other ingredients, under the widely varying global climatic conditions. To evaluate the model's robustness and veracity for application under such worldwide differences in operations, material systems, climates and environmental conditions, it is essential to employ the data generated worldwide. Accordingly in this chapter, worldwide data compiled from various literatures, were employed for evaluation of the NCP model. The data employed were obtained from independent sources worldwide, representing all the three (3) major geographical climate zones comprising the temperate, tropical and subtropical regions.

In the subsequent sections, the NCP model is evaluated using field-based experimental data. All the data employed were those taken from the various literature sources, wherein concrete samples had been cast then exposed outdoors to undergo natural carbonation under local weather conditions. The samples were then tested for natural carbonation progression as reported in the various literatures. Also given for each data set is a description of the relevant data characteristics that were considered essential to modelling.

4.4 HUY VU ET AL.'S (2019) DATA WORLDWIDE

4.4.1 EXPERIMENTAL SET-UP OF HUY VU ET AL.'S (2019)
CARBONATION EXPERIMENT

Huy Vu et al. (2019) reported the natural carbonation data of an investigation conducted on concretes exposed outdoors in five (5) cities and countries of Lyon (France), Austin (USA), Changsha (China), Chennai (India) and Fredericton (Canada). The concrete samples were prepared in France, then distributed by shipping to the other four (4) worldwide city locations. Upon arrival at the various destinations, the 100 × 100 × 400 mm prism samples were exposed outdoors to undergo natural carbonation under the local environmental conditions. The concretes comprised 45 mixtures of 0.45 to 0.65 w/cm's, cured at normal temperatures by wrapping the samples with a wet cloth, followed by sealing them in plastic bags, before shipping to the various destinations. Due to the various shipping time durations taken for the prisms to reach their respective destinations, the samples were subjected to different ages of prolonged curing varying from 7 to 163 days. Data of samples that had been cured for only one (1) day then exposed outdoors, were excluded from the present analysis since such short curing durations do not represent actual construction practice.

The concrete mixtures employed in the data (Huy Vu et al., 2019), had been prepared using a wide variety of composite cement types containing different proportions of the various SCMs specified in EN 197-1 (2000). These mixtures were made with or contained: clinker cement (CEM I), 11% limestone (CEM IIA-LL), 30% limestone (CEM IIB-LL), 15% blast-furnace slag (CEM IIA-S), 30% blast-furnace slag (CEM IIB-S), 50% blast-furnace slag (CEM IIIA), 15% fly ash (CEM IIA-V), 30% fly ash (CEM IIB-V), 50% fly ash, 13% or 15% pozzolan (CEM IIA-P), 30% pozzolan (CEM IIB-P), 20% blast-furnace slag + 20% fly ash (CEM VB), 20% blast-furnace slag + 20% pozzolan (CEM VB), 20% fly ash + 20% pozzolan (CEM IVB). Evidently, the data comprised most varieties of cement composites covering all the standard CEM I to V designations. The concretes that had been exposed outdoors in the five (5) site locations at different worldwide cities, were tested for carbonation depth at the ages of 1, 2, 3 and 5 years.

4.4.2 CLIMATES AT THE OUTDOOR EXPOSURE SITE LOCATIONS WORLDWIDE

The outdoor exposure sites were located in five (5) cities of the different countries representing various climates across worldwide geographical regions, as shown in

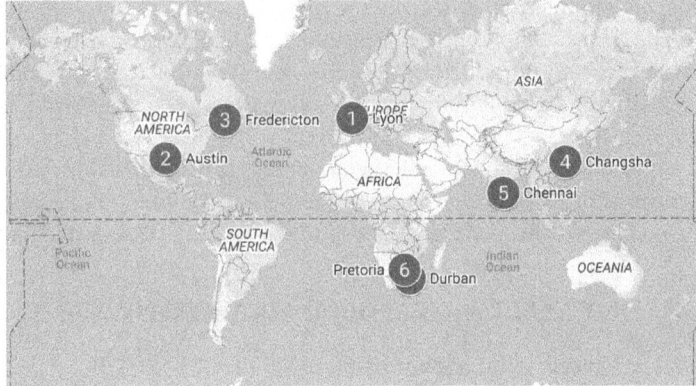

FIGURE 4.1   Worldwide locations of the outdoor exposure sites.

Figure 4.1. It can be seen in Figure 4.2 and Table 4.2, that the worldwide climates and weather conditions at the exposure locations, varied vastly across the different continents (*https://en.climate-data.org*). The climate variations stretched across the predominantly inhabited geographical regions worldwide from the extreme temperate very cold winter weather of Fredericton in Canada, to the equatorial wet/dry hot tropical conditions of Chennai in India.

Figure 4.2a–e shows that the cities of Lyon, Austin and Fredericton all have a similar temperate climate pattern, with rainfall generally falling uniformly throughout the year, while winter season occurs over the six (6) month period from October to April the following year. But the RH and temperature levels of the three (3) cities, are different. The warmest of the cities is Austin with temperatures of 10 to 30°C compared to the colder 3.2 to 21°C of Lyon and the freezy −9.0 to 20°C of Fredericton. The colder cities of Lyon and Fredericton both have higher RH levels of 60% to 83% and 67% to 76%, respectively, relative to the lower 55% to 66% of the warmer Austin. While Changsha also receives rainfall throughout the year, a peak season of very heavy rains occurs over the six (6) month period from March to September. Its RH range of 60% to 80% throughout the year, is similar to that of Lyon. Also, Changsha's winter is a short season of four (4) to five (5) months from November to March the following year. The city's temperature range of 6 to 29°C is between those of Lyon and Austin.

Meanwhile, the city of Chennai has a different climate pattern of alternating dry and wet seasons. The two-season equatorial hot tropical climate of Chennai comprises the dry months from January to April, followed by a wet season of heavy rains for rest of the year. Of all the five (5) site locations, Chennai experiences the hottest temperatures, being 25 to 30°C all year round.

It is evident from the foregoing that the worldwide climate extremes and variations employed in Huy Vu et al.'s (2019) data are quite severe, and were thus interestingly used to intensely evaluate the model's robustness, adaptability and universal applicability when predicting the natural carbonation tendencies of concrete structures located at various parts of the world.

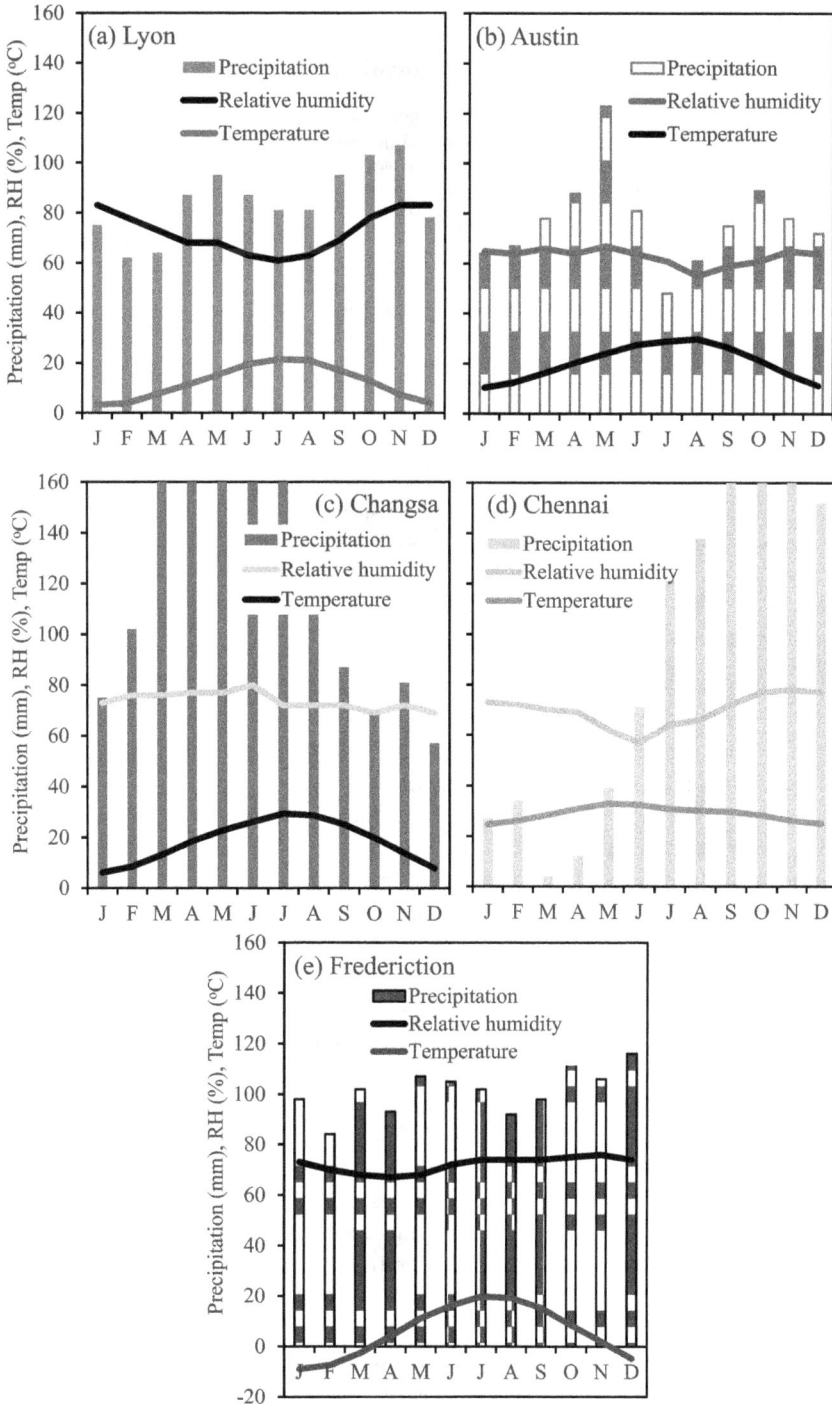

FIGURE 4.2 Climates at the site locations worldwide (a) Lyon, (b) Austin, (c) Changsha, (d) Chennai and (e) Fredericton (https://en.climate-data.org).

TABLE 4.2

Climate conditions of the various city locations.

Location	Average annual temperature (°C)	Average annual relative humidity, RH (%)	Climate
Lyon, FRANCE	13.0	68.9	Temperate with cold winter and humid warm summer climate.
Chennai, INDIA	28.7	70.9	Equatorial wet/dry, hot and humid.
Austin, USA	20.2	66.1	Temperate with cold winter and humid hot summer climate.
Changsha, CHINA	18.2	73.4	Temperate with cold winter and humid hot summer climate.
Fredericton, CANADA	6.4	73.2	Temperate with very cold winter and humid warm summer climate.

Source: (data from Huy Vu et al., 2019; https://en.climate-data.org).

4.4.3 COMPARISON OF THE MODEL'S PREDICTIONS WITH ACTUAL CARBONATION RESULTS MEASURED WORLDWIDE

Huy Vu et al.'s (2019) natural carbonation data described in Section 4.4.1, were employed to compare measured results versus the model's predictions, for concretes exposed outdoors at the urban sites located in each of the five (5) countries comprising France, India, USA, Canada and China.

Using Huy Vu et al.'s (2019) data, the NCP model described in Chapter 3 was rigorously evaluated to examine its robustness and veracity when employed to predict natural carbonation under the different worldwide climates. Given in Table 4.2 are values of the environmental parameter inputs used for each exposure site location. CO_2 concentrations at the different sites were not reported in the data, hence the historical record of global emissions was instead used (IPCC, 2013). Accordingly, the CO_2 concentration of about 400 ppm observed in urban settings during the year A.D. 2000, was employed in the model-based calculations.

For purposes of the present validation study, EN 197-1 (2000) cement designations were assigned to composites used in the data, based on types and proportions of SCMs incorporated into the concrete mixtures (Huy Vu et al., 2019). Evidently, the concretes exposed at each of the sites worldwide were made with cement varieties covering all the standard CEM I to V composite designations. Using Equations (3.2) and (3.5) of the NCP model (Chapter 3), correction adjustments were determined to account for effects of climate factors comprising RH and temperature at each exposure site location.

4.4.3.1 Model's predictions versus actual measured results

Figure 4.3a–e shows graphs comparing the model's predictions with actual carbonation results, for *sheltered* samples exposed outdoors at the different site locations worldwide. The scatter plots show that the model gave realistic predictions of the actual natural carbonation that occurred under the different climates worldwide. This performance of the model is indicative of its robustness, given the wide range

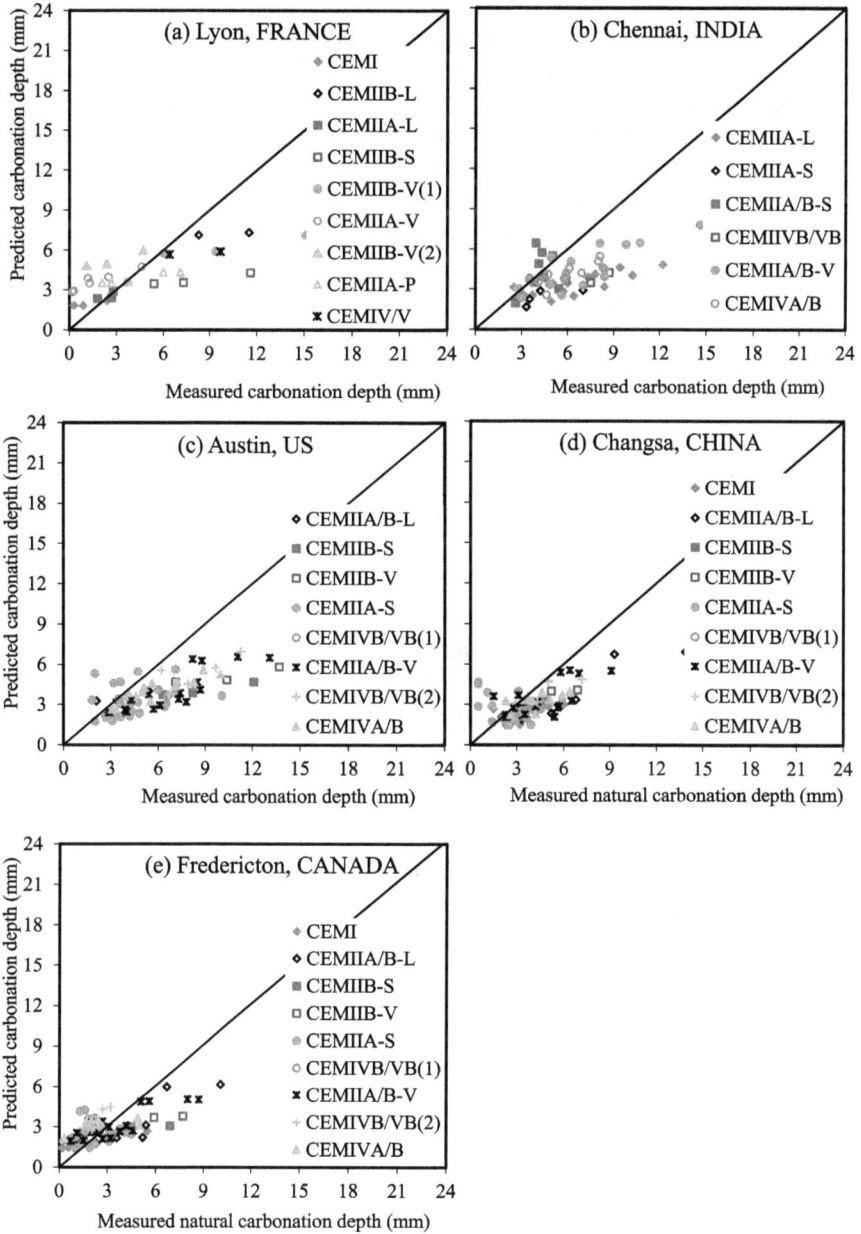

FIGURE 4.3 Scatter plots comparing the natural carbonation prediction results of NCP model versus the actual measured values for *sheltered* concretes exposed at the different worldwide site locations of (a) Lyon in France, (b) Chennai in India, (c) Austin in USA, (d) Changsha in China and (e) Fredericton in Canada.

of different cement composites used in the various concrete mixtures, along with the extreme climate variations of the worldwide data employed. It can be seen that for each of the exposure sites, the data points generally fall along the line of equality, regardless of the varied age of curing utilized along with different durations of concrete exposure to natural carbonation at the various sites. The observed data scatter is characteristic of the random tendency of natural phenomena. Variability of measured or predicted results is expectedly higher, owing to the wide range of cement composites used in concrete mixtures, along with the extreme climate variations to which the samples were exposed. It may be noted that the widely varied and prolonged curing durations reported in the data (Section 4.4.1), are not consistent with use of the model's f_{c28} parameter, an aspect that most likely affected the general pattern of predicted values seen in Figure 4.3a–e. It would have been preferable to use f_{cbn} values determined at the age of site exposure of samples, however, no such data were available.

It may be recalled that the cities of Lyon, Austin and Changsha all have similar cold temperate climate conditions with winter temperatures of about 3 to 12°C, while their summer temperatures are 20 to 30°C. Evidently, the results for all three (3) sites (Figure 4.3a, c, d) showed a similar pattern, although their samples had been subjected to different curing ages. It may be recalled that for Lyon site, the 100 × 100 × 400 mm prisms were exposed immediately after 28 days of curing, whereas the samples for all other locations were shipped while undergoing prolonged curing, prior to their outdoor exposure upon reaching the various site destinations (Section 4.4.1).

Fredericton site represents the cold temperate climate typical of weather in the Northern hemisphere. This site location has a very cold snowy winter weather with an average annual temperature of 6.4°C and falling as low as −20°C, while the maximum summer temperatures may rise to about 25°C. Interestingly, the graph for Fredericton site (Figure 4.3e) also gave a similar pattern of results as those of the other sites (Figure 4.3a–d).

The observed realistic prediction results discussed in the foregoing, underscore the model's ability to effectively determine corrective adjustments that dynamically account for effects of RH and temperature changes at different site locations worldwide. Chennai's climate is a two-season, all-year wet/dry weather that is vastly different from the four (4) season (winter, autumn, summer and spring) weather conditions of all the other site locations. Indeed, Chennai's equatorial climate exhibiting the all-year annual temperatures of 20 to 35°C, is representative of the typically wet/dry tropical conditions at geographical regions near equator across different continents. Evidently, the NCP model performed effectively for concretes exposed to the tropical conditions (Figure 4.3b), as it similarly performed realistically for mixtures exposed to carbonation under the snowy cold weather (Figure 4.3e), or under the intermediate climate variations (Figure 4.3a, c, d) falling between the aforementioned two (2) extremes.

4.4.3.2   Residuals and error statistics

Residuals between the predicted and the actual carbonation depths, are shown in Figure 4.4. Heteroscedasticity of the residuals is evident, with all the plotted graphs showing 'fanning out'. This observed tendency of residuals to 'fan out' as

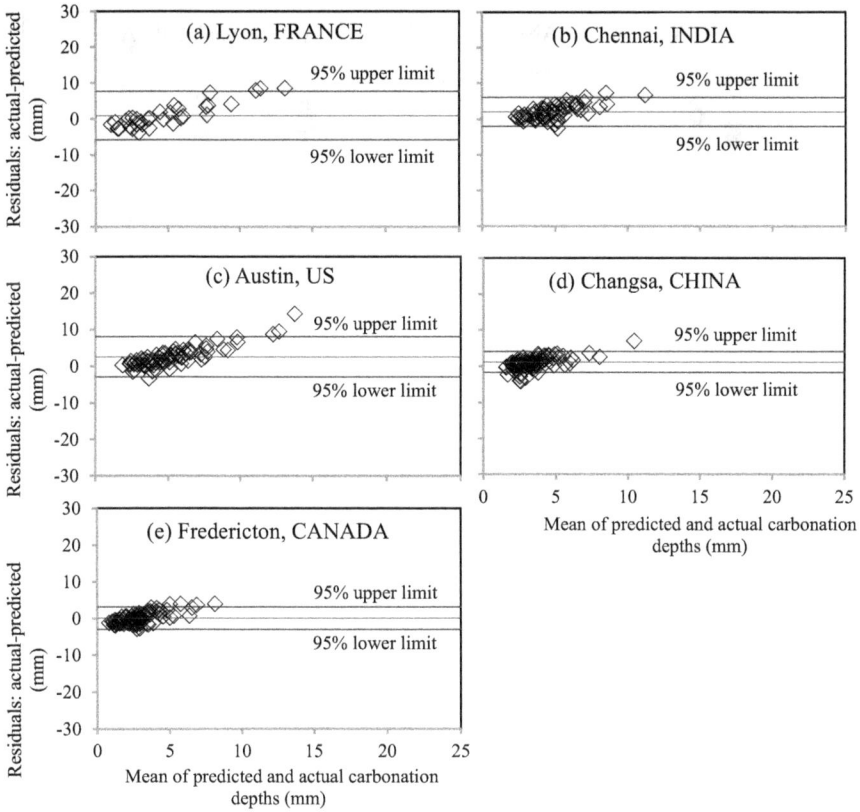

FIGURE 4.4 Plots of residuals against carbonation depth.

carbonation depth values increased, is a characteristic feature also exhibited by various other natural phenomena of concrete including creep and shrinkage (ACI 209, 1997; Gardner and Lockman, 2001; Bazant and Panula, 1978; Bazant and Baweja, 1995), concrete strength and workability (Naghizadeh and Ekolu, 2019). It appears that the 'fanning out' of residuals as herein observed, may be attributed to both concrete *quality bias* and carbonation *duration bias*. At any stage of carbonation progression, lower quality concretes give higher carbonation depths, relative to those of higher quality mixtures. Typically, as the material's quality reduces, it exhibits correspondingly higher variability of the measured property. As such, the lower quality concretes which accordingly gave higher carbonation depths, exhibited correspondingly higher residuals, leading to the 'fanning out' heteroscedasticity seen in Figure 4.4, thereby depicting *quality bias*. Similar 'fanning out' heteroscedasticity is also seen in Figure 4.5, but this scenario shows higher residuals with corresponding increase in duration of concrete exposure to natural carbonation, an observation which depicts *duration bias*.

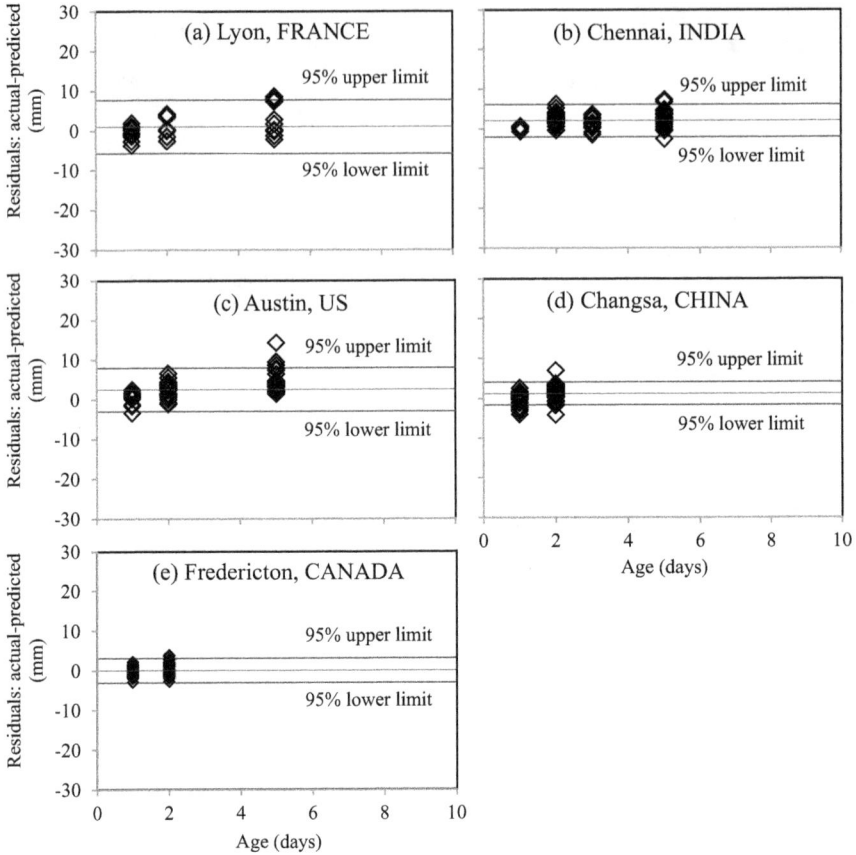

FIGURE 4.5 Plots of residuals against age.

While heteroscedasticity is not indicative of the model's prediction veracity, it could underscore an accuracy limit beyond which prediction results may not be of much use. From analysis of the present data, carbonation depth of about 6 mm seemed particularly interesting as the model tended to under predict results that were higher than this value. It appears that acceleration of carbonation progression occurs in mixtures containing > 30% content of the various SCMs other than slag. It is on this basis that use of the model is presently limited to composites containing ≤ 30% common SCMs or ≤ 50% slag (Section 3.4).

Table 4.3 gives error statistics of the model's predictions. The error analysis conducted was based on three (3) statistical indicators comprising the ratio of mean measured value (MV) to mean predicted value (PV), root mean square of error (RMS) and coefficient of variation of error (CV). It can be seen in Table 4.3 that the MV/PV ratio exhibited a fairly large spread of values above or below 1.0, an observation that can be attributed to the extensive range of cement composites used in mixtures, along with the extreme climate variations to which the samples were exposed. Moreover, the prolonged curing employed in the data is not consistent with use of the

TABLE 4.3
Error statistics.

Site location	MV/PV*	RMS	Coefficient of variation of error, CV (%)
Lyon, FRANCE	1.22	3.46	64.0
Chennai, INDIA	1.53	2.88	48.2
Austin, USA	1.62	3.72	56.5
Changsha, CHINA	1.25	1.86	48.9
Frederiction, CANADA	1.01	1.53	53.6

Note: *MV is the measured value, PV is the predicted value, RMS is the root mean square of error.

model's f_{c28} parameter. Nonetheless, most MP/PV values were between about 1.0 to 1.2 (Table 4.3), which indicates realistic performance by the model. Also, most CV values (Table 4.3) were within the established range of 30% to 50% (Ekolu, 2018), which is consistent with findings from the associated studies (Ekolu, 2020a, 2020b; Ekolu and Solomon, 2021). It may be noted that the model's CV values obtained as shown in Table 4.3, are comparable to the accuracy CV level of 20% to 50% (Bazant and Baweja, 1995; Lifecon, 2003), being the established range of values for code-type prediction models comprising ACI 209 (1997), CEB-FIP (1990), *fib* 34 (2006) and Model B3 (Bazant and Panula, 1978).

4.5 DATA FROM VAN DIJK (1987) AND BOUTEK (1994), PRETORIA AND DURBAN, SOUTH AFRICA

4.5.1 OUTDOOR CARBONATION EXPERIMENTS

The experiments comprised two (2) data sets of natural carbonation investigations conducted in the South African cities of Pretoria and Durban. Experimental details of the data sets (Kruger, 2003) are subsequently described.

- Van Dijk (1987) reported an early investigation done at the National Building Research Institute (NBRI, South Africa, https://www.csir.co.za/), wherein OPC concretes of 0.43, 0.65 and 0.87 w/cm's with or without 30% fly ash (CEM I, CEM IIB-V), were exposed to natural carbonation. Cubes of 100 mm size were cast and cured in a fog room for three (3) days, then exposed outdoors under Pretoria weather conditions. Carbonation depth values were measured at the ages of 364 and 728 days.
- BOUTEK (1994) was also a study involving the natural carbonation of 20, 40 and 55 MPa concretes made with or without 30% fly ash (CEM I, CEM IIB-V). Concrete prisms of 100 x 100 x 500 mm size were cast and subjected to a curing regime comprising: the storage of samples in a fog room for one (1) day, following which they were wrapped with polyethylene sheet, then

further kept at 50% RH/23°C for six (6) days and stored in laboratory air for three (3) weeks. After these 28 days of curing, the samples were exposed outdoors at sites in Pretoria city or at Durban seacoast. Carbonation depth measurements were done at the ages of 16 and 60 months.

4.5.2   PRETORIA AND DURBAN CLIMATES

The subtropical weather of Pretoria comprises a dry winter season occurring at mid-year over a six (6) month period from April to September, followed by a rainy season during the other six (6) months of the year. Durban is a coastal city, which explains its high RH of 76.1%, while the inland city of Pretoria has lower RH of 40% to 62%. Durban is consistently warm with temperatures of 17 to 23°C. Pretoria has similar temperatures as Durban, except in the winter months of May to August during which temperatures fall lower within 12 to 15°C.

4.5.3   COMPARISON OF THE MODEL'S PREDICTIONS WITH ACTUAL CARBONATION RESULTS

The South African data reported by Van Dijk (1987) and by BOUTEK (1994) were generated during the period when CO_2 concentration at urban centres was about 350 ppm (IPCC, 2013). As such, the [CO_2] level of 350 ppm was employed in the model-based calculations for the samples exposed at the inland site of Pretoria city. A lower CO_2 concentration of 250 ppm was assumed for the coastal site of Durban city, considering that oceans absorb about 30% of atmospheric CO_2 emissions.

Figure 4.6a shows a scatter plot comparing the model's predictions with actual natural carbonation values measured for concretes exposed at Pretoria and Durban site locations. Again, it can be seen that the model's predictions are generally correct, with data points falling along the line of equality, as similarly observed in Figure 4.3a–e for Huy Vu et al.'s (2019) data. It may be recalled that Durban is a coastal city while Pretoria is an inland urban centre. The standard cement types CEM I and CEM II/B-V, were used in the concretes exposed to natural carbonation at both sites. Since the concrete samples exposed at Durban site were near the seacoast, chloride ingress expectedly occurred along with carbonation. From the observations seen in Figure 4.6a, it is evident that chloride ingress did not significantly influence natural carbonation, considering that the model gave realistic predictions for the concretes exposed to the marine environment.

Plotted graphs of residuals versus carbonation depth as given in Figure 4.6b, exhibit the 'fanning out' heteroscedasticity, as similarly observed for Huy Vu et al.'s (2019) data shown in Figure 4.4. However, the plot of residuals versus age as seen in Figure 4.6c, did not show heteroscedasticity. The model predictions for carbonation of the concretes exposed at Pretoria and Durban sites, gave CV values of 28.7% and 44.8%, respectively, both of which fall within the model's accuracy CV range of 30% to 50%, as similarly observed for Huy Vu et al.'s data (2019) (Section 4.4.3).

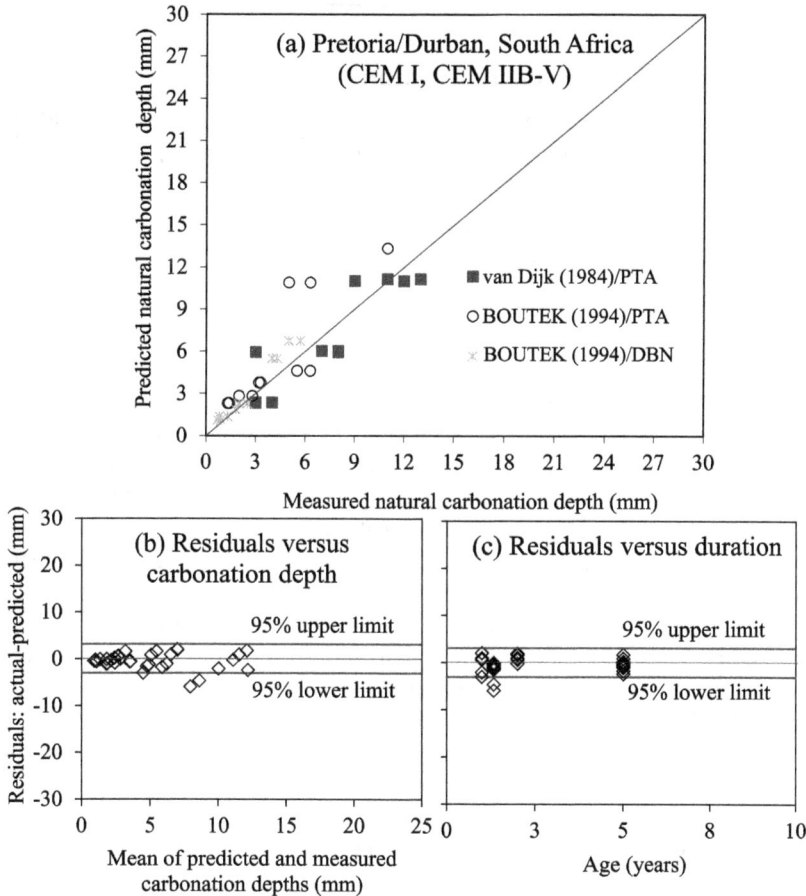

FIGURE 4.6 Carbonation predictions for concretes exposed outdoors at sites in Pretoria (PTA) and Durban (DBN) cities (a) scatter plot comparing the model's predictions with actual results, (b) residuals versus carbonation depth and (c) residuals versus duration of concrete carbonation.

4.6 YUNUSA'S (2014) DATA, JOHANNESBURG, SOUTH AFRICA

The data reported by Yunusa (2014) were employed to evaluate the model's prediction performance owing to the relevance therein particularly focused on concretes containing SCMs or SCM concretes. The data comprised carbonation results of concretes that had been made with CEM I 52.5 cement containing 0% or 10% silica fume (SF), 30% fly ash (FA) or 50% ground granulated blast-furnace slag (SG), along with results of concretes made with the standard CEM V/A (S-V) 32.5N cement. In the experiment, SCMs were incorporated into concrete mixtures by blending the pozzolans with CEM I. Table 4.4 gives the equivalent standard cement designations identified for each of the blended cement types used in

TABLE 4.4
Cement types used in the concrete mixtures.

Cement type used	CEM I	10SF*	30FA	50SG	50–60NP/FA
Equivalent standard cement (EN 197-1, 2000)	CEM I	CEM II/A	CEM II/B, CEM IV/A	CEM III/A, CEM IV/B	CEM V/A

Note: *SF is silica fume, FA is fly ash, SG is ground granulated blast-furnace slag, NP is natural pozzolan.

mixtures. It can be seen that the experiment included nearly all standard CEM I to V cement types typically used for structural concretes.

The data comprised concrete mixtures made at water/cementitious (w/cm) ratios of 0.4, 0.5, 0.6, 0.75 and cementitious contents of 300, 350, 400, 450 kg/m^3. The 19 mm granite stone and granite crusher sand, were used in all mixtures. Concrete cubes of 100 mm size were cast and moist-cured in water at room temperature for 3, 7 and 28 days, following which compressive strengths were determined while other samples were exposed outdoors to undergo natural carbonation.

Four (4) contiguous cube surfaces were epoxy-coated, leaving two (2) uncoated opposite surfaces for unidirectional CO_2 ingress. The cube samples were then exposed to three (3) different outdoor environments comprising: the flat roof of a medium-storey building, basement parking garage of a multi-storey building and a site beneath a pedestrian bridge spanning over a busy highway. The cubes stored at the roof were *unsheltered*, while those kept underneath the bridge overpass were *sheltered* from rain. Since the basement parking was enclosed, this exposure condition is an indoor environment and was therefore not considered in the present analysis (Section 3.5.1). Carbonation measurements were done at the ages of 6, 12, 18 and 24 months.

The annual microclimate conditions at the outdoor *sheltered* and *unsheltered* exposure sites, were 230 to 250 ppm CO_2/50% RH/19°C. During the months of precipitation from October to April, the measured average RH values at the outdoor *sheltered* and *unsheltered* sites were 58% RH and 59% RH, respectively. These measured values are consistent with the macroclimate of Johannesburg city, being an annual average of 59% RH/22°C.

4.6.1 COMPARISON OF THE MODEL'S PREDICTIONS
WITH ACTUAL CARBONATION RESULTS

Carbonation values predicted using the NCP model are given in Figure 4.7 for each of the respective cement types, and compared with corresponding experimental results. The data for the *sheltered* and *unsheltered* exposure sites are both plotted on the same graph of a given cement type, but they have been distinguished using the filled and non-filled markers, respectively. Only the data from the two (2) outdoor *sheltered* and *unsheltered* exposure sites were employed. No attempt was made to employ data from the indoor site for carbonation prediction, since the NCP model is presently only applicable to outdoor exposure conditions (Section 3.5.1). Altogether 496 data points were used.

It can be seen in scatter plots of Figure 4.7 that in each case, data points fall along the line of equality, which shows good agreement between the predicted and

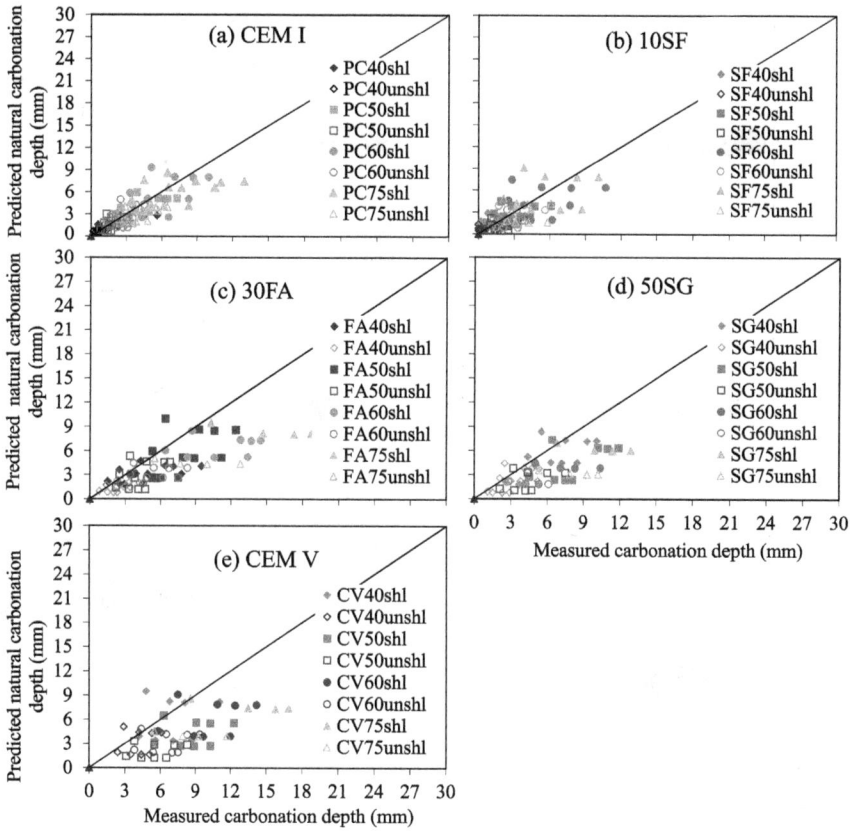

FIGURE 4.7 Comparison of predicted carbonation depths versus actual measured results: *shl* is sheltered, *unshl* is unsheltered.

measured carbonation depths. This observation is evident for each of the five (5) cement types. Compressive strengths of CEM I concretes ranged from 23 to 76 MPa, while 10SF concretes gave high strengths of up to 91 MPa. The 30FA and 50SG concretes were mostly low and medium strength mixtures of up to 60 MPa, while CEM V concretes had low strengths of 20 to 30 MPa range, except the two (2) mixtures made at 0.4 and 0.5 w/cm which attained the 28-day strength of about 40 MPa.

Figure 4.7 shows that CEM I and 10SF concretes that were mainly moderate to high strength mixtures, gave lower carbonation depths relative to those of 30FA, 50SG and CEM V concretes. The maximum carbonation depths predicted for CEM I and 10SF concretes were about 9 mm, while higher maximum values of around 12 mm were predicted for the other cement types containing FA and SG.

The plotted graphs of residuals as shown in Figure 4.8, exhibit the extent of over-prediction and under-prediction made by the model, relative to measured values. Again, the higher quality CEM I and 10SF concretes of medium and high strengths along with 30FA and 50SG mixtures, showed the least residuals between

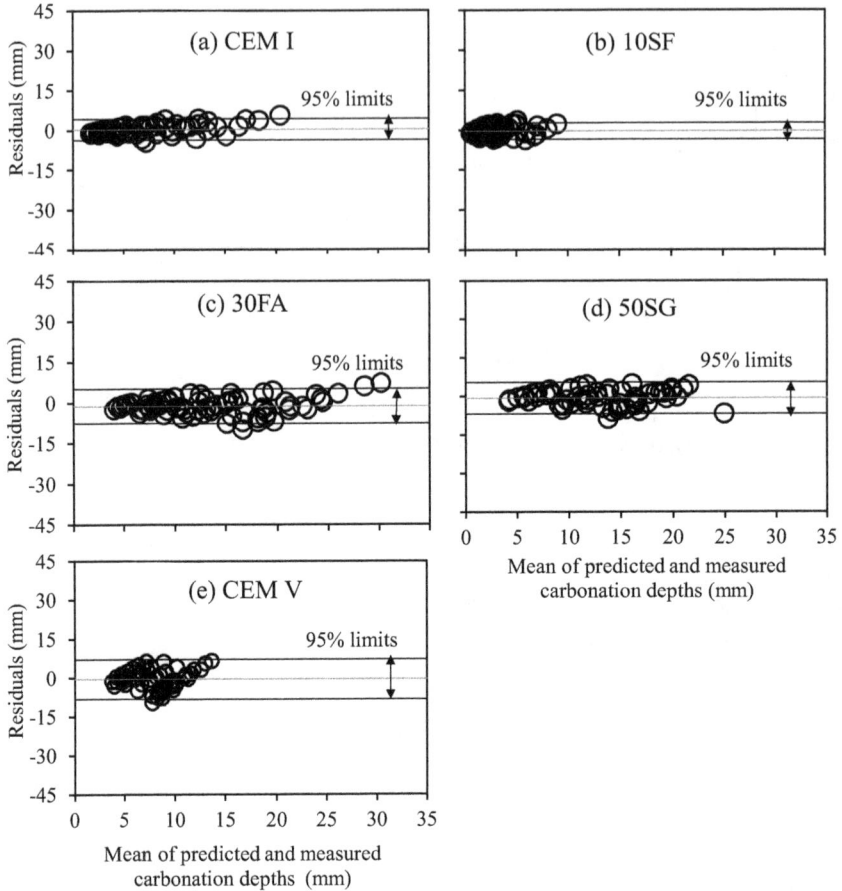

FIGURE 4.8 Residuals plotted against the mean of predicted and measured carbonation depths.

the predicted and measured values. Also included in the graphs of Figure 4.8, are the mean residuals and 95% limits of the confidence interval. It can be seen that the residuals fall within the 95% confidence interval, implying that the model gave correct predictions of actual carbonation depth.

4.6.2 ERROR ANALYSIS

Statistical indicators of the model's prediction accuracy are given Table 4.5. They comprise the ratio of mean measured value (MV) to mean predicted value (PV), root mean square of error (RMS) and coefficient of variation (CV) of errors. It can be seen that MV/PV values are close to 1.0 in all cases. Clearly, the model's CV values of 35% to 41% are similar to those of the creep and shrinkage code-type models,

TABLE 4.5
Statistical error indicators.

	Concrete mixtures									
	CEM I		10SF		30FA		50SG		CEM V	
Statistical indicators	MV*	PV	MV	PV	MV	PV	MV	PV	MV	PV
Mean (mm)	3.7	4.3	3.2	3.4	7.8	7.5	8.6	6.9	9.2	8.2
Mean MV/PV		0.9		1.0		1.0		1.3		1.1
RMS		1.5		1.4		2.9		2.4		2.9
CV (%)		35.3		41.3		39.0		34.5		35.9

Note: *MV is the measured value, PV is the predicted value, CV is the coefficient of variation of errors.

which typically give CV accuracy values of 20% to 50% (Bazant and Baweja, 1995; Lifecon, 2003).

4.7 BOUZOUBAA ET AL.'S (2010) DATA, OTTAWA, CANADA

4.7.1 OUTDOOR EXPERIMENTAL STUDY

Bouzoubaa et al.'s (2010) data comprised 25, 35 and 45 MPa concrete mixtures made with ASTM Type 1 cement containing 0%, 25%, 35% and 50% fly ash. The equivalent ASTM C595 (2020) designation of blended cements used in the mixtures, was Type 1P(X) where X is SCM proportion in the blended cement. The fly ash materials used were obtained from two (2) Canadian sources of Lingon and Sundance. Large prisms of size $100 \times 100 \times 400$ mm were cast and cured in water for seven (7) days, then exposed outdoors *unsheltered*. Carbonation depths were measured after four (4) years.

Ottawa's climate comprises the annual average RH and temperature of 71.2% and 6.9°C, respectively. The temperate climate is earmarked by cold winter with six (6) months of sub-zero temperatures from November to April of the following year.

4.7.2 COMPARISON OF THE MODEL'S PREDICTIONS WITH ACTUAL CARBONATION RESULTS MEASURED IN OTTAWA

The comparison of measured with predicted carbonation rates, is shown in Figure 4.9a for the *unsheltered* concretes exposed outdoors in Ottawa. Evidently, the model's predictions are realistic as depicted by data points falling along the line of equality. The CV value of 38.5% obtained is consistent with those determined based on other data sources analyzed earlier (Sections 4.4 to 4.6).

Also shown in Figure 4.9b are residuals between the model's predictions and measured results. Evidently, the residuals fall along the zero-line and within the 95% confidence limits, showing that the model gave correct predictions. The observed

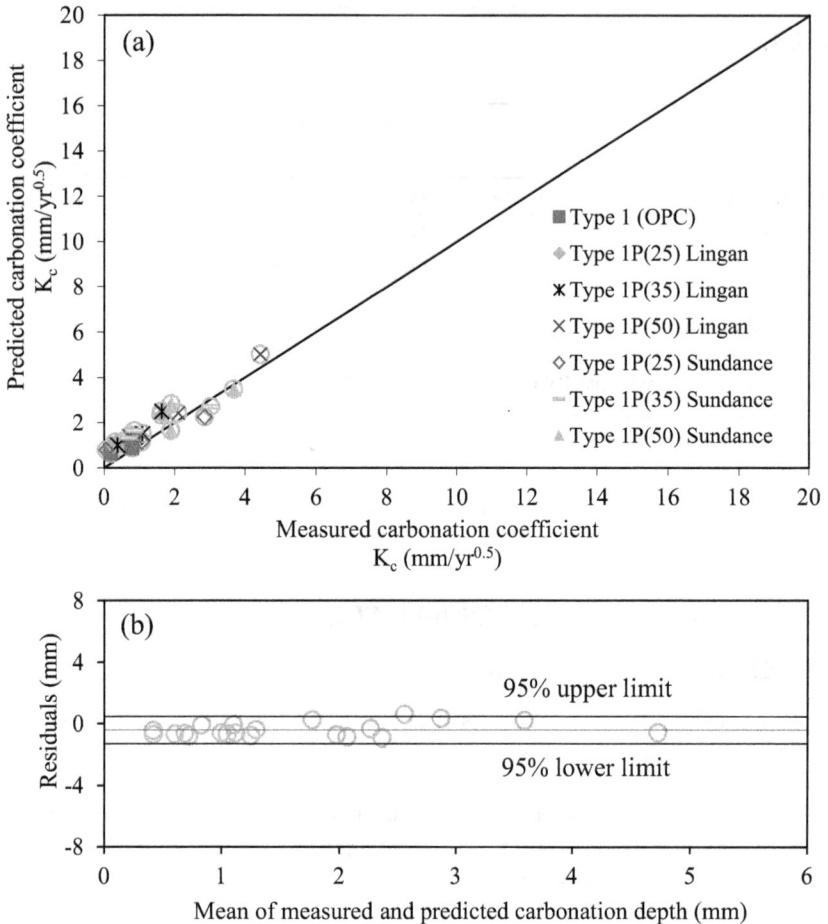

FIGURE 4.9 Concrete carbonation in Ottawa, Canada showing (a) comparison of predicted versus measured carbonation rates and (b) residuals.

'fanning-out' heteroskedasticity exhibited by the residuals (Figure 4.9b), is attributed to higher variability as concrete quality decreased at lower strengths, in turn giving higher carbonation results.

4.8 BUCHER ET AL.'S (2017) DATA, TOULOUSE, FRANCE

4.8.1 OUTDOOR EXPERIMENTAL INVESTIGATION

The data reported by Bucher et al. (2017) comprised carbonation results of various 50 MPa concretes prepared using CEM I to V cement types containing different SCM varieties consisting of fly ash (FA), limestone filler (LL), ggbs and metakaolin (MK). Ternary blends of the SCMs with CEM I comprised 22% FA + 22% ggbs, 16% LL + 15% MK and 15% FA + 15% MK to produce standard cements of the

TABLE 4.6

Cementitious systems used in the concrete mixtures.

Cement designation	Cement grade	w/cm	Composite system	Blend	SCM total content (%)
CEM I	52.5	0.6	CEM I normal		0
CEM IP MES	52.5	0.6	CEM I low $C_3A =2.5\%$		0
CEM II/A-LL	42.5	0.53	16% LL	Binary	16
CEM II/A-V	42.5	0.53	15% FA	Binary	15
CEM III/A	52.5	0.6	62% ggbs	Binary	62
CEM V/A	42.5	0.53	22% FA+22% ggbs	Ternary	44
CEM I15	52.5	0.6	15% MK	Binary	15
CEM I20	52.5	0.6	20% MK	Binary	20
CEM I25	52.5	0.6	25% MK	Binary	25
CEM II/A-LL15	42.5	0.53	16% LL+15% MK	Ternary	31
CEM II/A-V15	42.5	0.53	15% FA + 15% MK	Ternary	30

designations CEM V/A, CEM II/A-LL15 and CEM II/A-V15, respectively. Table 4.6 gives the full range of cementitious systems employed. Cube samples of 100 mm size were cast and cured in water for various ages of up to 28 days. After curing, the cube samples were exposed outdoors, then carbonation depths were measured after one (1) year and after two (2) years. Compressive strength values were also determined at the ages of 3, 5, 28 and 90 days.

The temperate climate of Toulouse city comprises six (6) months of winter from November to April the following year, while rainfall occurs significantly throughout the year, giving a high annual RH of 71.5% and average temperature of 13.8°C.

4.8.2 COMPARISON OF THE MODEL'S PREDICTIONS WITH ACTUAL CARBONATION RESULTS MEASURED IN TOULOUSE

Figure 4.10 shows the plotted graph of measured versus predicted carbonation depths of the concretes that had been stored outdoors under *sheltered* exposure conditions. It can be seen that the model gave correct carbonation predictions for the various concretes, as indicated by data values falling along the line of equality. This realistic performance of the model is consistent with those observed for all prior evaluations conducted throughout this chapter. The CV of 25.2% obtained, highlights the model's good prediction accuracy as consistently found for the various foregone evaluations discussed in Sections 4.4 to 4.7.

Figure 4.10b shows that the residuals fall along the zero-line and within the 95% confidence limits, thereby affirming that the model gave realistic predictions. The

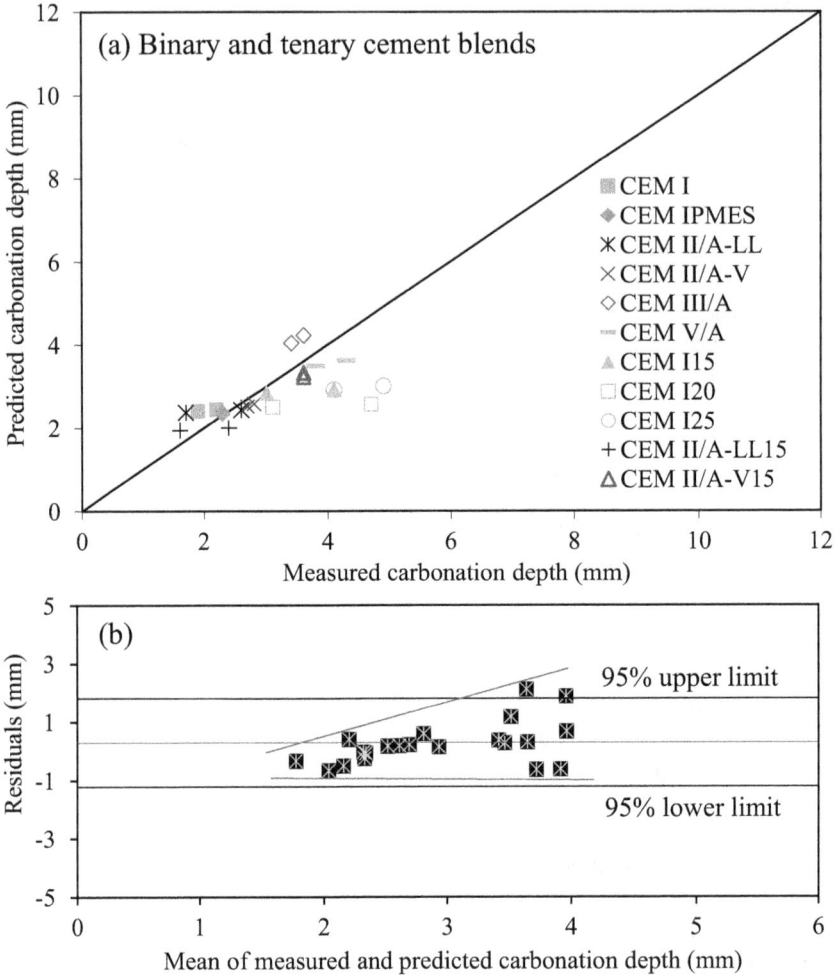

FIGURE 4.10 Concrete carbonation in Toulouse, France showing (a) comparison of predicted versus measured carbonation depths and (b) residuals.

'fanning-out' heteroskedasticity can be seen becoming more pronounced as carbonation depths increased, an observation attributed to higher variability of lower quality concretes, as already explained in Sections 4.4.3.2 and 4.7.2.

4.9 CONCLUSION

This chapter focused on validation of the NCP model using various experimental data of outdoor natural carbonation results drawn from worldwide literature sources. The model's robustness and veracity were tested under the widely varied range of factors known to significantly influence carbonation of concrete structures worldwide. These factors include different cementitious systems comprising standard cement composites and blended cements used in different parts of the world, for different structural concretes of varied strengths, under various curing regimes, extreme exposure conditions and global climates, among others. The foregone Sections 4.4 to 4.8 focused on the model's validation using a wide range of data sets of different characteristics. Interestingly, the NCP model gave realistic predictions under all the various scenarios.

In Section 4.4, the NCP model was validated using worldwide data comprising natural carbonation of concretes in five (5) cities and countries of Lyon (France), Austin (USA), Changsha (China), Chennai (India) and Fredericton (Canada). Accordingly, the model's performance was evaluated under the influence of extreme environmental conditions and world climates (Huy Vu et al., 2019). In Section 4.5, South African data were employed to evaluate performance of the model for prediction of natural carbonation under the inland Pretoria weather and under Durban's coastal marine environment (Van Dijk, 1987; BOUTEK, 1994). The NCP model was also successfully validated in Section 4.6, using a large set of 496 data points based on the South African outdoor natural carbonation study conducted in Johannesburg. The concrete mixtures employed had been prepared using all major CEM I to V standard cements of EN 197-1 (2000), then subjected to natural carbonation under both *sheltered* and *unsheltered* outdoor exposure conditions (Yunusa, 2014).

The model's validation given in Section 4.7 was based on Canadian data of natural carbonation of concretes in Ottawa, wherein North American cements of ASTM Type 1 and Type 1P(X) were employed to prepare mixtures (Bouzoubaa et al., 2010). The French data of natural carbonation in Toulouse, was also similarly employed for the model's validation, as discussed in Section 4.8 (Bucher et al., 2017). A key aspect of the French data was the extensive range of binary and ternary cement blends used to produce eleven (11) different cement composites of CEM I to V varieties.

It can be seen in the foregoing that validations done in the various Sections 4.4 to 4.8, all confirm the model's robustness and veracity consistently giving realistic carbonation predictions, having been rigorously tested under the different system extremes. The validations also found that the model's prediction accuracy consistently gave CV values of 20% to 50%, which are similar to those of code-type models.

REFERENCES

ACI 209 (1997) American Concrete Institute (ACI) Committee 209, Subcommittee II. *Prediction of creep, shrinkage and temperature effects in concrete structures*, Report ACI 209 R92, (re-approved 1997).

ASTM C595 (2020) *Standard specification for blended hydraulic cements*, ASTM International, West Conshohocken, PA.

Bazant Z.P. and Baweja S. (1995) Justification and refinements of Model B3 for concrete creep and shrinkage, 1. Statistics and sensitivity, *Materials and Structures*, 28, 415–430.

Bazant Z.P. and Panula L. (1978) Practical prediction of time-dependent deformations of concrete, *Materials and Structures (RILEM, Paris): Part I Shrinkage*, 11, 307–316.

BOUTEK (1994) *Corrosion of steel in and durability properties of OPC, FA and GGBS concretes, Pretoria: Division of Building Technology (BOUTEK)*, CSIR (Contract Report No. BB 078 5600 5671 to Ash Resources).

Bouzoubaa N., Bilodeau A., Tamtsia B. and Foo S. (2010) Carbonation of fly ash concrete: laboratory and field data, *Canadian Journal of Civil Engineering*, 37(12), 1535–1549, https://doi.org/10.1139/L10–081

Bucher R., Diederich P., Escadeillas G. and Cyr M. (2017) Service life of metakaolin-based concrete exposed to carbonation: comparison with blended cement containing fly ash, blast furnace slag and limestone filler, *Cement and Concrete Research*, 99, 18–29.

CEB (1997) New approach to durability design-an example for carbonation induced corrosion, In: *Bulletin*, vol. 238 (P. Schiessl, Ed.), Comite´ Euro-International du Be´ton, Lausanne.

CEB-FIP Model Code (1990) *Design code 1994*, Thomas Telford, London.

CSIR (1999) *Final report on an investigation into the influence of fly ash on the durability of concrete*, by W.R. Barker, Division of Building and Construction Technology, CSIRPretoria, South Africa, 18p. Also, Ash Resources Also, Ash Resources (pty) Ltd, PO Box 3017, Randburg 2125.

Ekolu S.O. (2018) Model for practical prediction of natural carbonation in reinforced concrete: part 1-formulation, *Cement and Concrete Composites*, 86, 40–56.

Ekolu S.O. (2020a) Model for natural carbonation prediction (NCP): practical application worldwide to real life functioning concrete structures, *Engineering Structures*, 224, 111126. http://doi.org/10.1016/j.engstruct.2020.111126

Ekolu S.O. (2020b) Implications of global CO_2 emissions on natural carbonation and service lifespan of concrete infrastructures – reliability analysis, *Cement and Concrete Composites*, 114, 103744, http://doi.org/10.1016/j.cemconcomp.2020.103744

Ekolu S.O. and Solomon F. (2021) A case study on practical prediction of natural carbonation for concretes containing supplementary cementitious materials, *KSCE Journal of Civil Engineering*, 26(3), 1163–1176, http://doi.org/10.1007/s12205-021-1770-6

EN 197-1 (2000) *Cement – part 1: composition, specifications and conformity criteria for common cements*, European Committee for Standardization, CEN, Management Centre, Rue de Stassart, 36 B-1050 Brussels, Belgium, 29p.

fib 34 (2006) Model code for service-life design, fib bulletin 34, Federation International du Beton, Lausanne, 1st Ed., 126p., ISBN: 978-2-88394-074-1.

Gardner N. and Lockman (2001) GL-2000, Design provisions for drying shrinkage and creep of normal strength concrete, *ACI Materials Journal*, 98, 159–167.

Huy Vu Q., Pham G., Chonier A., Brouard E., Rathnarajan S., Pillai R., Gettu R., Santhanam M., Aguayo F., Folliard K.J., Thomas M.D., Moffat T., Shi C. and Sarnot A. (2019) Impact of different climates on the resistance of concrete to natural carbonation, *Construction and Building Materials*, 216, 450–467.

IPCC (2013) Annex II climate system scenario tables. In: *Climate Change 2013: the physical science basis, contribution of working group I to the fifth assessment report of the Intergovernmental Panel on Climate Change (IPCC)* (T.F. Stocker, D. Qin, G.-K. Plattner, M. Tignor, S.K. Allen, J. Boschung, A. Nauels, Y. Xia, V. Bex and P.M. Midgley, Eds.), Cambridge University Press, Cambridge.

Kruger J.E. (2003) *South African fly ash: a cement extender*, The South African Coal Ash Association, P.O. Box 3981, Cresta, 2118, RSA, ISBN 0-958-45349-7, 219p.

Lifecon (2003) *Deliverable D 3.2 service life models: life cycle management of concrete infrastructures for improved sustainability*, Final Report by Dipl.-Ing. Sascha Lay, Technical Research Centre of Finland (VTT), 169p.

Naghizadeh A. and Ekolu S.O. (2019) Method for comprehensive mix design of fly ash geopolymer mortars, *Construction and Building Materials*, 202, 704–717.

Papadakis V.G., Fardis M.N. and Vayenas C.G. (1992) Effect of composition, environmental factors and cement-lime mortar on concrete carbonation, *Materials and Structures*, 25, 293–304.

Parrot L.J. (1987) *A review of carbonation in reinforced concrete*, Cement and Concrete Association, BRE, 42p.

Solomatine D., See L. and Abrahart R. (2009) Data-driven modelling: concepts, approaches and experiences. In: *Practical hydroinformatics*, Water Science and Technology Library, Vol. 68, 14p, https://doi.org/10.1007/978-3-540-79881-1_2

Van Dijk J. (1987) The influence of FA on the properties of hardened concrete cured under various conditions, Ash – a valuable resource. In: *First International Symposium*, S387, Pretoria, February, Vol. 2.

Yunusa A.A. (2014) The effects of materials and micro-climate variations on predictions of carbonation rate in reinforced concrete in the inland environment, *PhD Thesis*, School of Civil Engineering & the Built Environment, University of the Witwatersrand, Johannesburg, RSA.

5 Validation of NCP model's application to real-life concrete structures

5.1 ENGINEERING MODELS

Practical engineering models typically possess three (3) basic characteristics comprising: veracity, operational practicality and cost-effectiveness. The foremost basic characteristic of a practical engineering model is its ability to veraciously depict or predict the phenomenon's progression. Carbonation along with other processes such as chloride ingress, water absorption, concrete creep and shrinkage, etc., are natural phenomena governed by laws of physical science and can be expressed as mathematical functions (ACI 209, 1992; CEB-FIP, 1990; Gardner and Lockman, 2001). Engineering models such as employed in structural design are typically analytical mathematical methods that are not necessarily computational but are sufficiently empirical to be executed with simple tools such as hand or manual calculators and basic computing. Obviously, such methods can be implemented using computer programming software packages. Academic programmes of engineering education typically strive to develop student understanding of fundamental principles. Based on this approach, a design engineer should be capable of employing analytical thinking so as to maintain full control of the design process with ability to independently check computerized outputs, using basic design tools.

To validate a model's veracity, it is essential to utilize data from experimental investigations that depict real-life scenarios or to utilize data from real-life engineering structures. Both approaches have been employed in this book to validate the NCP model. Chapter 4 presented the model's validation using various data sets that were based on experimental set-ups depicting real-life scenarios. The present Chapter 5 concerns validation of the model using carbonation data of real-life existing concrete structures.

5.2 NATURAL CARBONATION DATA OF REAL-LIFE CONCRETE STRUCTURES

This chapter is based on data of real-life functioning concrete structures in service. Again, the NCP model is validated for robustness and veracity using data from worldwide sources. Accordingly, the data examined the model's behaviour under

DOI: 10.1201/9781003645399-5

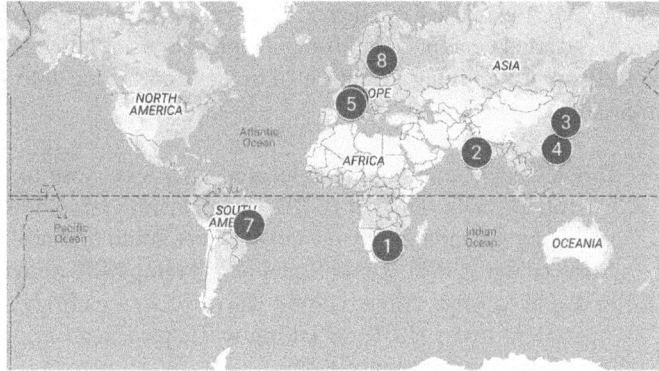

FIGURE 5.1 Locations of the real-life concrete structures globally (Ekolu, 2020).

extremes that were not limited to particular localized conditions, but that instead covered a vast range of exposure environments of the temperate, tropical and subtropical climate regions. Moreover, the data employed were taken from both old and new engineering structures that had been in service for different ages. It may also be added that all data of the real-life structures employed in the validation study, are independent results from the various literatures.

The structures considered in the validation study comprised existing reinforced concrete buildings and bridges located in various regions worldwide, as seen in Figure 5.1 (Ekolu, 2020). The major urban locations of the existing concrete structures are Johannesburg (South Africa), Bhopal (India), Seoul (South Korea), Taipei (Taiwan, China), Turin (Italy), Blenio (Switzerland), Brasilia (Brazil) and Tallin (Estonia). Subsequently given is the NCP's model performance upon application to predict natural carbonation progression in each of the existing concrete structures. The model's predictions are then compared with the actual measured values of ongoing natural carbonation in the structures.

5.3 PREDICTION OF THE NATURAL CARBONATION OF FIFTEEN (15) EXISTING JOHANNESBURG BUILDINGS AND BRIDGES

5.3.1 DESCRIPTION OF THE JOHANNESBURG STRUCTURES

Lampacher (2000) reported carbonation and compressive strength measurements done on fifteen (15) real-life concrete structures comprising bridges and buildings built between 1962 and 1973 in the greater Johannesburg city. The oldest of structures was the Yale Telescope building built in 1922, while the youngest structure was the then 19-year old Booysens Road on/off ramp bridge constructed in 1973.

Cores were reportedly extracted from structural elements, then tested for compressive strength and carbonation depth. The other tests reportedly done were durability indexes comprising water sorptivity, oxygen permeability and chloride conductivity. The compressive strength values determined varied from 29.1 to 61.2 MPa. It was reported in the literature (Lampacher, 2000) that ordinary Portland cement (OPC) was used in all the concrete mixtures, except for three (3) bridges wherein blended cements containing slag had been employed. Evidently, the study involved a wide-ranging selection of structures falling into three (3) categories comprising:

- Urban bridges and buildings located within or close to the Johannesburg Central Business District (CBD). The existing structures along with their years of construction are: the Goch Street South bridge (1965), Goch Street North bridge (1966), Rissik Street on/off ramp M2E bridge (1968), M2 E/W bridge at Loveday Street (1968), Harrow/Saratoga bridge (1962), the Ponte Apartment tower (1970) in Hillbrow, Empire Road bridge (1968) and the Yale Telescope building (1922).
- Suburban bridges located at city outskirts or Johannesburg suburbs, being the Diepsloot 2nd bridge (1972) and Diepsloot 3rd bridge (1972).
- Highway bridges subsequently listed along with their years of construction: St. Andrews Road bridge (1968), N4 bridge No. 2597 Witbank (1966), 1st Avenue bridge Parktown (1971), Corlett Drive bridge (1972) and Booysen's Road on/off ramp bridge (1973).

It can be seen that the concrete structures are located across the wide geographical spatial triad of Johannesburg-Pretoria-Witbank, comprising the three (3) major cities engulfing the high pollution belt of Eskom's coal-based electricity generating power plants located in Mpumalanga province.

Figures 5.2 and 5.3 are among the fifteen (15) Johannesburg bridges and buildings mentioned in the foregoing. The oldest structure was the Yale Telescope building

FIGURE 5.2   The Yale Telescope building (1922).

FIGURE 5.3 Empire Road bridge (1968).

(Figure 5.2) that was 70 years at the time of testing. The building's main structural components were reinforced concrete columns and walls made with cement/sand block cladding. Cores were extracted from the columns then used for conducting the various tests. It was reported that the aged building structure had undergone corrosion repairs in 1993.

The Empire Road bridge shown in Figure 5.3 is a multi-span two-way deck system connecting the N1 highway North with N1 highway South, which in turn joins the double-decker carriageway leading south past Johannesburg CBD. The bridge was built in 1968 with high strength 60 MPa concrete. At the time of testing, the structure was 24 years of age. Cores for carbonation testing were extracted from the bridge's massive piers.

About six (6) km south of Johannesburg CBD is the Booysen's Road on/off ramp bridge constructed in 1973 with moderate strength 40 MPa concrete. This single-span bridge is both an underpass and an overpass for the two-way traffic of N1 highway joining from or off-ramping to Booysen's road. The rest of other bridges had similar design features, concrete strengths and age categories, as those described in the foregoing. Meanwhile, the iconic Ponte Apartment tower built in 1970 with high strength 50 MPa concrete, is Johannesburg's cylindrical skyscraper of 55 storeys. The building's reinforced concrete columns at the basement, were core-drilled to obtain samples used to conduct the carbonation and concrete strength tests.

5.3.2 Johannesburg climate

The ambient climate of Johannesburg is subtropical weather with average annual temperature of 16.3°C along with 59.3% RH. The city's low annual precipitation of 543 mm is consistent with its moderate RH. Rainfall occurs during summer for the five (5) month period from December to April the following year. The other seven (7) months during the seasons of autumn (March to April), winter (May to July) and spring (August to November), are generally a period of dry spell with little to no rainfall.

5.3.3 Comparison of the model's predictions with measured carbonation of Johannesburg structures

The NCP model was employed to predict values of carbonation coefficient for each of the fifteen (15) concrete structures aged 20 to 70 years at the time of testing. The predicted values have been plotted against measured results as shown in Figure 5.4a, giving carbonation rates of about 1.5 to 5.0 mm/yr$^{0.5}$, depending on the concrete strength. As expected, the structures constructed with high strength concretes > 50 MPa, exhibited the lowest carbonation rates. Indeed, the Empire Road bridge built with 61.2 MPa concrete, Ponte Apartment tower constructed with 52.7 MPa concrete and Diepsloot 3rd bridge built with 54.4 MPa concrete, all gave low carbonation rates of 2.32, 1.40 and 2.18 mm/yr$^{0.5}$, respectively. Similarly, the St. Andrews Road bridge constructed with high strength concrete of 53.4 MPa gave the low carbonation rate of 1.18 mm/yr$^{0.5}$. Meanwhile, structures that had been constructed using low concrete strengths < 30 MPa, gave the highest carbonation coefficients. Indeed, the Harrow/Saratoga bridge and M2 E/W Loveday Street bridge built using 29.1 MPa and 29.4 MPa concretes, respectively, exhibited the highest rates of carbonation, being 4.65 and 5.1 mm/yr$^{0.5}$, respectively. All the other structures had been built with concretes of medium strengths ranging from 30 MPa to under 50 MPa, and they gave intermediate carbonation rates of 2 to 6 mm/yr$^{0.5}$.

It can be seen in Figure 5.4a that the data points fall along the line of equality, thereby indicating statistical equality between the predicted and measured rates of natural carbonation. These observations affirm the model's realistic predictions. Figure 5.4b shows a graph of residuals between the predicted and measured carbonation coefficients. Evidence of heteroscedasticity of the residuals can be seen, showing 'fanning out' as carbonation rates increased, a feature associated with increase in variability as concrete quality decreases (Chapter 4, Sections 4.4.3.2 and 4.7.2). The model's CV for all fifteen (15) concrete structures was determined to be 24%, indicating good accuracy of the model's predictions.

5.4 PREDICTION OF THE NATURAL CARBONATION OF TEN (10) EXISTING BHOPAL STRUCTURES

5.4.1 Description of the Bhopal structures

Rizvi et al. (2017) measured natural carbonation of existing concrete structures in India's city of Bhopal, while highlighting the ongoing high growth of urban population and infrastructure, as typical of most developing countries. The associated

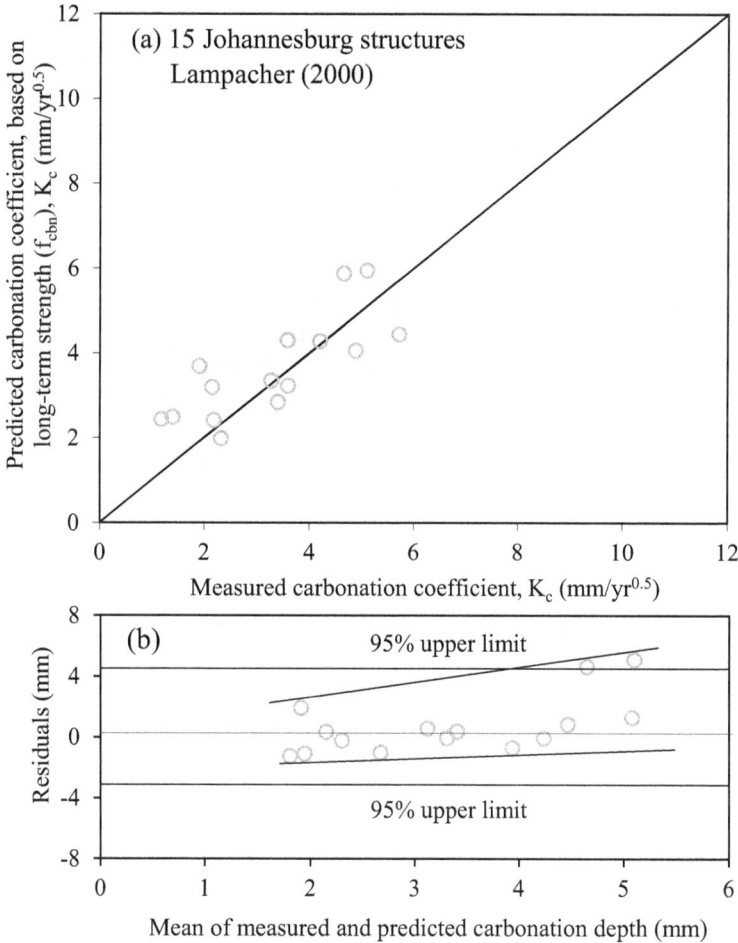

FIGURE 5.4 Existing fifteen (15) Johannesburg bridges and buildings (a) predicted versus measured carbonation coefficients and (b) residuals.

increase in vehicular traffic volume along with operation of large manufacturing industries needed to serve industrial and socio-economic needs of society, majorly contribute to atmospheric $[CO_2]$ rise. The measured data reported in the literature (Rizvi et al., 2017) for each structure comprised compressive strength, carbonation depth and concrete cover, along with age of the structure.

5.4.2 BHOPAL'S CLIMATE

The tropical wet/dry climate in Bhopal, comprises warm to hot temperatures throughout the year. The average monthly temperatures range from a minimum of 21°C to a maximum of 38°C, giving an annual average of 28.3°C. Bhopal city has a unique weather pattern consisting of a dry hot spell for most of the year, except during the four (4) months of July to October when short bursts of heavy rainfall occur. Consequently, the city's annual RH is very low, being merely 41.8%. Indeed, Bhopal's heavy rainfall is so intense over the four-month period, that its average annual precipitation of 2055 mm is among the highest in the world. It may be added that such heavy rainfall over a short-time duration, often causes flash flooding and its associated adverse effects.

5.4.3 COMPARISON OF THE MODEL'S PREDICTIONS WITH MEASURED NATURAL CARBONATION OF BHOPAL STRUCTURES

The geographical location of Bhopal represents interesting climate conditions of hot tropical weather extremes comprising low RH and warm hot temperatures. Of the twenty-five (25) concrete structures presented in the literature (Rizvi et al., 2017), only ten (10) of them met the model's condition wherein concrete strength values exceeded 20 MPa (Section 3.4, Eq. 3.9b), hence all others were discarded from the analysis. The environmental parameters employed in the analysis were CO_2 concentration of 400 ppm, 41.8% RH, temperature of 28.3°C and *sheltered* outdoor exposure condition, while the cement used was taken to be OPC. The ten (10) structures that satisfied all conditions for application of the model (Section 3.5.1), mostly had low to moderate concrete strength values of 20 MPa and 30 MPa.

A graph comparing the predicted natural carbonation depths with actual measured values, is shown in Figure 5.5a. Evidently, the model's predictions are realistic, with data values falling along the equality line, as also earlier observed for the Johannesburg structures (Section 5.3.3). Residuals between the predicted and actual measured values of natural carbonation, are also shown in Figure 5.5b, wherein limits of 95% confidence interval are included. Unsurprisingly, the residuals showed no heteroskedasticity, which is expected since the concretes used in the structures were of similar quality and similar strength values, being 20 to 30 MPa. The 'fanning out' would be expected had the concretes used in the structures been of widely varied low to high strengths such as the case for Johannesburg structures (Section 5.3.3).

As already mentioned, the concretes used in Bhopal structures were of similar strengths, implying that their structural responses would also be expectedly similar. Evidently, the measured carbonation rates obtained were similar, comprising 6 to 8 mm/yr$^{0.5}$ for all the structures. The model's CV of 32% obtained from error analysis, shows good prediction accuracy falling within the range of 20% to 50%, which is the typical accuracy level of code-type models (Bazant and Baweja, 1995a; Lifecon, 2003).

5.5 PREDICTION OF THE NATURAL CARBONATION OF TWO (2) TAIPEI BRIDGES

5.5.1 DESCRIPTION OF THE TAIPEI CONCRETE BRIDGE STRUCTURES

Two existing bridges in Taipei, comprising the Chorng-ching viaduct built in 1971 and Wann-fwu bridge constructed in 1987, were investigated by Liang et al. (2013) during which data of concrete properties and serviceability conditions, were acquired. The existing bridge structures are located in the chloride-laden environment. Bridge details including layout drawings, and sizes of the various structural components, are given in Liang et al. (2013). The various structural elements comprising piers,

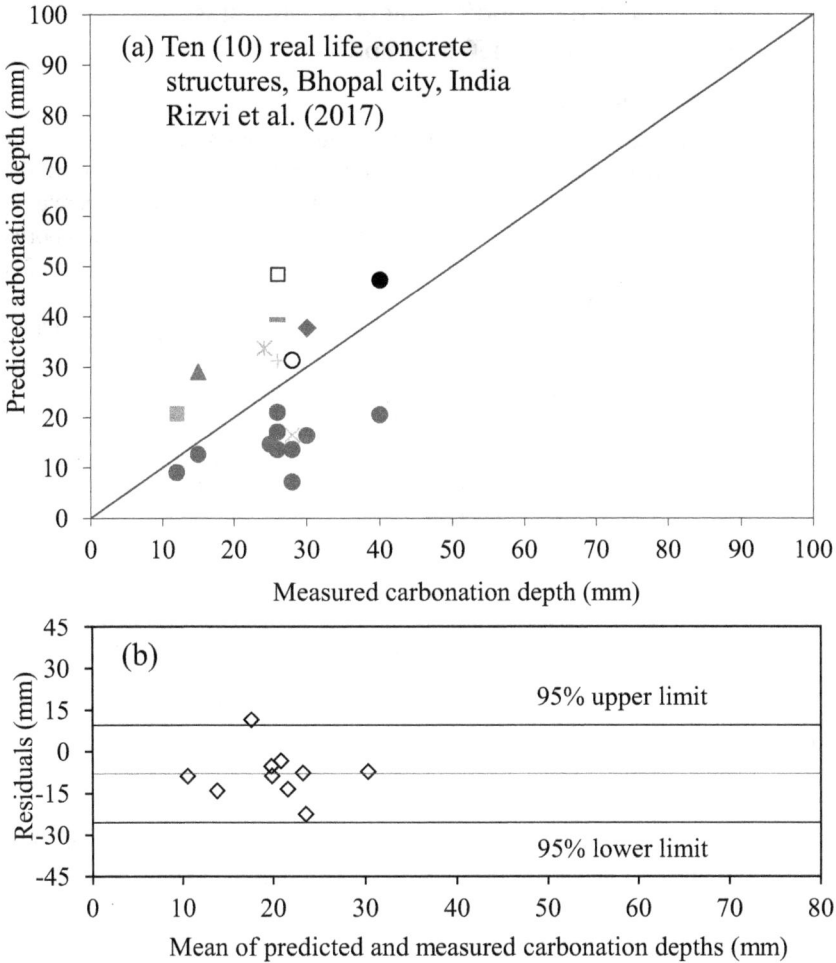

FIGURE 5.5 Existing ten (10) Bhopal concrete structures (a) predicted versus measured carbonation depths and (b) residuals.

girders, bridge deck, cap beam, abutments and retaining walls, were tested to obtain compressive strength results, carbonation depths and cover depths. Tests were done at twenty-one (21) different locations for the Chorng-ching viaduct and twenty (20) locations for Wann-fwu bridge.

5.5.2 TAIPEI'S CLIMATE

Taipei's weather is a warm to hot tropical climate with two (2) seasons consisting of wet and dry conditions, while temperatures range from 16 to 32°C throughout the year. Taipei receives heavy rainfall during the four (4) months of June to September, followed by moderate rains from October to February of the following year, prior to the dry season from March to May. Taipei's high annual precipitation of 1763 mm is responsible for its elevated average RH of 70.2%, while the city's average annual temperature is 26°C.

5.5.3 COMPARISON OF THE MODEL'S PREDICTIONS WITH MEASURED NATURAL CARBONATION OF TAIPEI BRIDGES

5.5.3.1 Chorng-ching viaduct

Results of the Chorng-ching viaduct, showed that its various structural elements had been built with concretes of different strength grades. The piers, abutments and retaining walls, were built with concretes of 20 to 30 MPa strength values, while concrete of 40 to 50 MPa was used to cast the girder. The measured carbonation rates were reportedly 2 to 6 $mm/yr^{0.5}$. The 17 data points that satisfied the model's conditions (Section 3.5.1), were employed in the analysis. The *sheltered* outdoor exposure condition was assumed for all test points of the various structural elements, while the $[CO_2]$ level of 350 ppm was used, being the average value from 1971 when the bridge was constructed to the year of testing (*https://sealevel.info/co2.html*). Ordinary Portland Cement (OPC) was assumed to have been used in the concretes, since no cement type had been reported in the literature. Employment of OPC in the analysis is consistent with concrete technology of the 1960's to 1990's when mainly clinker cements and Portland blast-furnace slag cement, were predominantly used in the construction industry.

Figure 5.6a shows comparison between the predicted and measured carbonation depth values for each test point. The data point of average values is also plotted on the same graph. It can be seen that the plotted data points fall reasonably along the line of equality. Also given in Figure 5.6b is the plotted graph of residuals between the predicted and measured values. Heteroskedasticity of the residuals is expectedly evident, with 'fanning out' representing increased variability depicted by concretes of relatively lower quality and strength.

5.5.3.2 Wann-fwu bridge

Natural carbonation of Wann-fwu bridge was similarly analyzed as in Figure 5.6 for Chorng-ching viaduct. The $[CO_2]$ of 370 ppm was used in the model, being approximately the average CO_2 concentration over the structure's lifespan from 1987 when the bridge was constructed to the time of testing. Moderate strength concretes of 30 to 40 MPa had been used to construct the bridge deck, beams and retaining walls. The carbonation rates of 2 to 7 $mm/yr^{0.5}$ measured for Wann-fwu bridge are similar to the reported 2 to 6 $mm/yr^{0.5}$ for Chorng-ching viaduct.

Figure 5.7a shows comparison of predicted versus the measured values of carbonation depths for each test point. Again, it can be seen that the model gave realistic predictions. Evidently, the data points fall along the line of equality. The residuals plotted in Figure 5.7b also showed 'fanning out' heteroskedasticity as similarly observed for the Chorng-ching viaduct.

5.6 PREDICTION OF THE NATURAL CARBONATION OF TWO (2) SWISS BUILDINGS IN BLENIO

5.6.1 DESCRIPTION OF THE CLASSROOM BUILDING AND GYMNASIUM

Located at Lugano in Blenio valley 580 m above sea level, are two (2) concrete structures comprising a classroom main building and a gymnasium built in 1979 and 1985, respectively (Teruzzi and Cadoni, 2003). Condition appraisal was

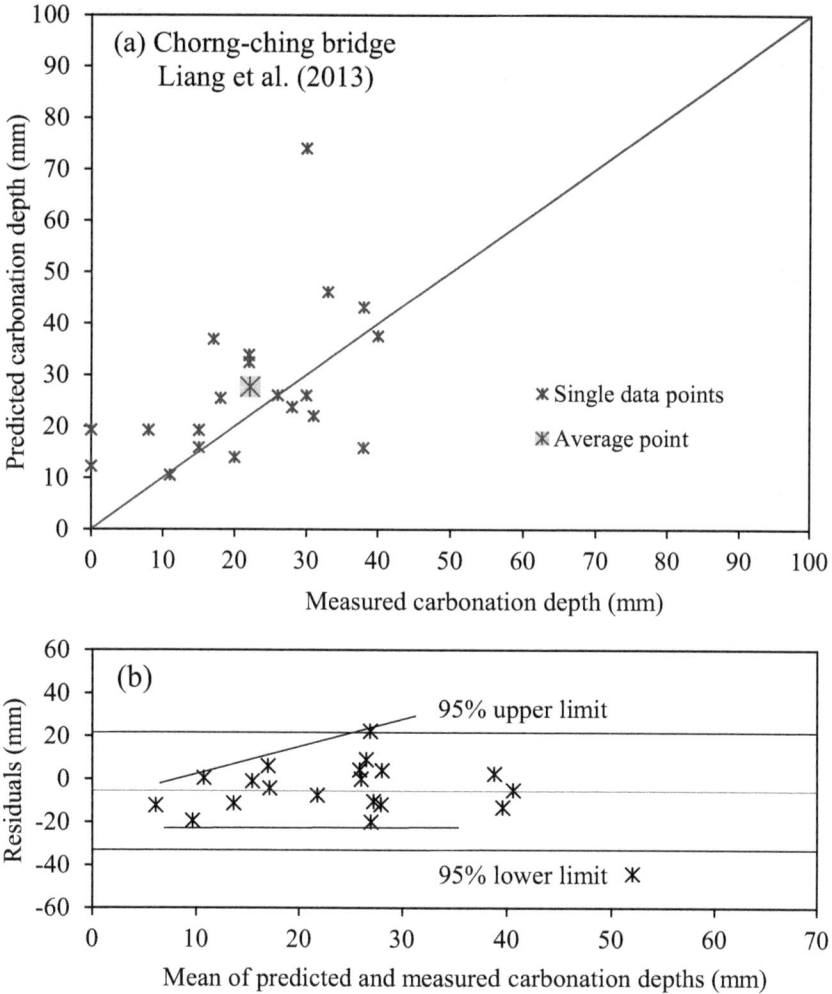

FIGURE 5.6 Chorng-ching viaduct in Taipei (a) predicted versus measured carbonation depths and (b) residuals.

reportedly done as part of infrastructure management undertaking. Structural elements of the buildings were identified during inspection, then cores were extracted from facades and tested to determine concrete strength, carbonation depth and cover depth.

5.6.2 Lugano's Blenio Swiss climate

Located in the Alpine mountains, Lugano region receives high rainfall most of the year, except in the winter months of December to March the following year, when precipitation is very low. The Alpine's high precipitation is responsible for the regions all year high average RH of 73.5%. Alpine temperatures range from 4 to 9°C during the four (4) winter months of December to March and 12 to 22°C for all the

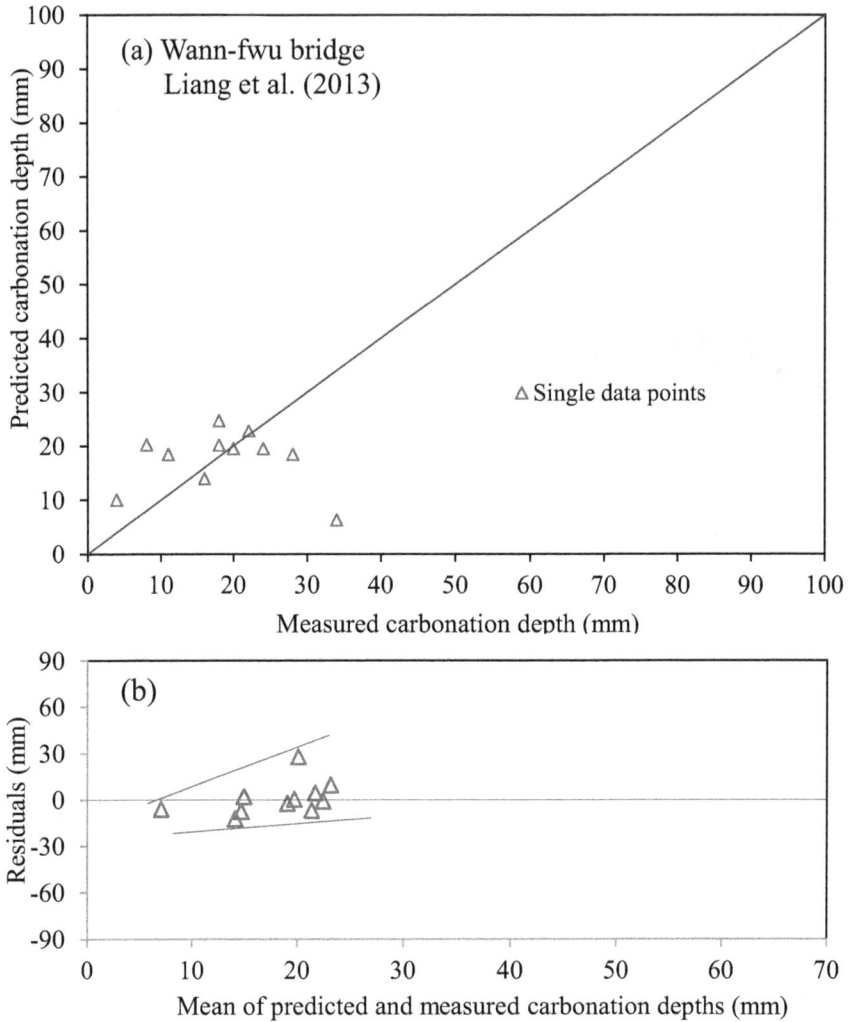

FIGURE 5.7 Wann-fwu bridge in Taipei (a) predicted versus measured carbonation depths and (b) residuals.

other months of the year. The region's annual climate comprises 705.5 mm precipitation, 73.5% RH and 14.5°C average temperature.

5.6.3 COMPARISON OF THE MODEL'S PREDICTIONS WITH MEASURED NATURAL CARBONATION OF SWISS BUILDINGS

Measured carbonation rates of the 17-year main building and 23-year gymnasium, were reportedly between 3 to 5 mm/yr$^{0.5}$. For carbonation prediction, the [CO_2] of 360 ppm was used, being the approximate average value from the time of construction to time of testing. Although the facades are typically exposed, they could have been effectively *sheltered* depending on the building's canopy and orientation to the direction of rainfall. Accordingly in this analysis, it was assumed that test points at the facades were *sheltered*. The model's predicted values can be seen in Figure 5.8a showing agreement with measured natural carbonation results, as indicated by the data points falling along the line of equality. Interestingly, the graph of residuals plotted in Figure 5.8b seemed to show convergence heteroskedasticity, which is a departure from the commonly observed 'fanning out' (Sections 5.3.3 and 5.5.3).

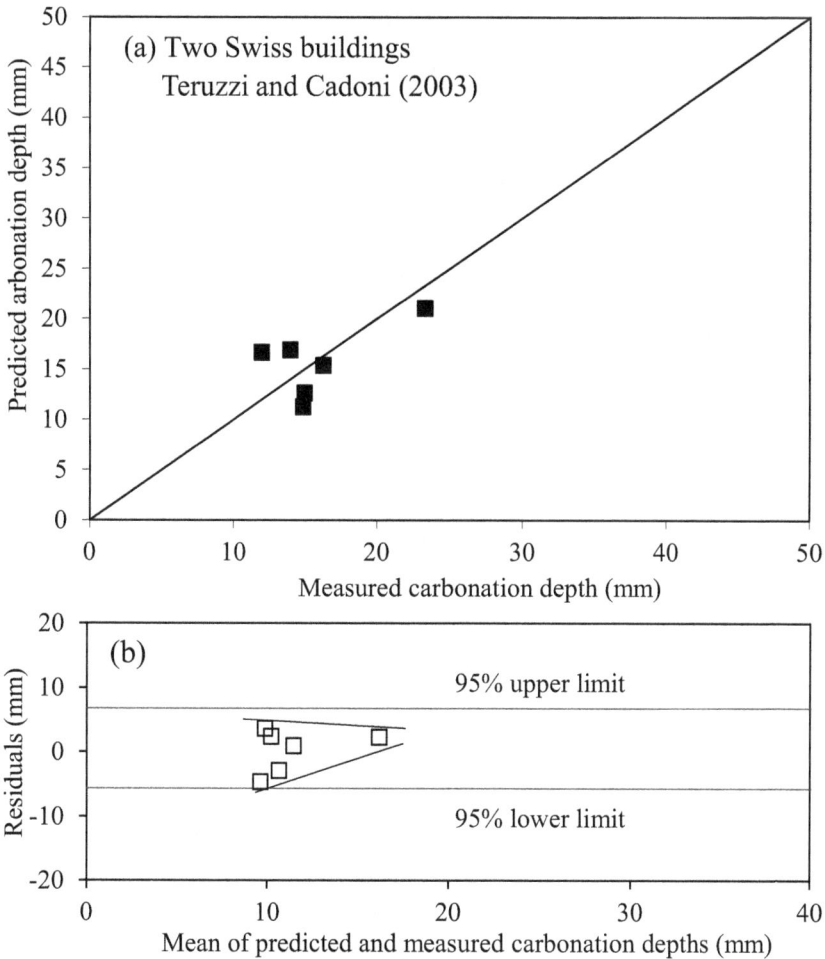

FIGURE 5.8 Two (2) Swiss buildings in Blenio, Lugano (a) predicted versus measured carbonation depths and (b) residuals.

5.7 PREDICTION OF THE NATURAL CARBONATION OF SIX (6) REINFORCED CONCRETE BRAZILIAN BUILDINGS

5.7.1 DESCRIPTION OF THE EXISTING REINFORCED CONCRETE BUILDINGS IN BRASILIA

Figueiredo and Nepomuceno (2005) investigated the natural carbonation of six (6) functional reinforced concrete buildings located in Brasilia city. The concrete structures comprised an 8-year old building in Aguas Claras, 22-year old Colina building, 25-year old Applied Studies School building at University of Brasilia, 33-year old NOVACAP building and the Monumental Axle building that was 35 years old. Six (6) columns of each building were core-drilled at locations of 1.0 to 1.5 m above ground level, to extract 70 dia. x 150 mm cylindrical concrete samples. The columns selected for coring, were those exposed to outdoor environment but *sheltered* from rain.

5.7.2 BRASILIA'S CLIMATE

The equatorial climate consisting of the two-season wet/dry warm tropical weather, occurs in Brasilia all year round, while annual temperatures are within 19 to 23°C. The annual 629.9 mm precipitation is attributed to the seven (7) month rainy period from October to April of the following year, while the subsequent five (5) month period from March to September is a dry spell. Expectedly, RH is low during the dry months of May to September while high during the rainy months of October to April. The average annual climate of Brasilia is 62.3% RH/22°C.

5.7.3 COMPARISON OF THE MODEL'S PREDICTIONS WITH MEASURED NATURAL CARBONATION OF BRAZILIAN BUILDINGS

The strength values of concretes used in the columns varied from 22 to 35 MPa, giving an average of 27 MPa for all the buildings. The measured carbonation coefficients were reportedly between 4.0 to 7.5 mm/yr$^{0.5}$. The [CO_2] level of about 350 ppm was used, being the average value between 1970 taken as the approximate year of construction to 2003 the presumed year of testing (*https://sealevel.info/co2.html*). OPC was assumed to have been used in all the concretes.

Figure 5.9 shows comparison of the predicted versus the measured values of natural carbonation in each of the reinforced concrete buildings. Again, it can be seen that the model gave good predictions with data values falling along the line of equality (Figure 5.9a). Also, the model's CV of 44.6% obtained, indicates good prediction accuracy similar to those of code-type models (Bazant and Baweja, 1995a, 1995b; Lifecon, 2003). The plotted graph of residuals seen in Figure 5.9b, shows 'fanning-out' heteroskedasticity, which is consistent with typical observations discussed in earlier sections of this chapter.

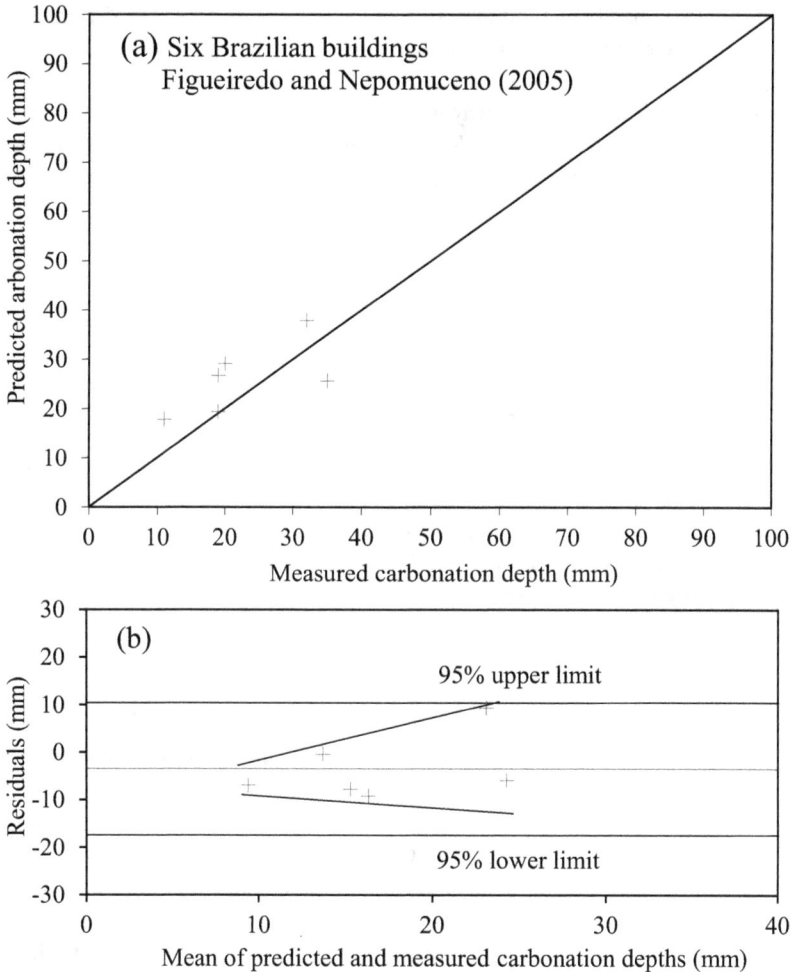

FIGURE 5.9 Six (6) reinforced concrete buildings in Brasilia (a) predicted versus measured carbonation depths and (b) residuals.

5.8 PREDICTION OF THE NATURAL CARBONATION OF ELEVEN (11) ESTONIAN BRIDGES

5.8.1 DESCRIPTION OF THE EXISTING HIGHWAY AND URBAN BRIDGES IN TALLIN

Liisma et al. (2017) investigated eleven (11) highway and urban bridges to evaluate natural carbonation progression. Ages of the bridges widely varied from 10 to 41 years. Carbonation depth values were determined along with *in situ* compressive strengths measured using the non-destructive technique (NDT) employing the rebound hammer test method. Measurements were done at heights of 0.3 to 0.5 m, or at 1.7 m above ground level. Four (4) different concrete strength grades comprising C25/30, C25/35,

C30/37 and C35/45 had been used in design of the various bridges. The different bridge structures along with their years of construction comprise: Kahala I (highway bridge, 1977), Kahala II (highway bridge, 1977), Kopli (urban bridge, 1976), Kuusalu I (highway bridge, 1989), Kuusalu II (highway bridge, 1989), Võidujooksu (urban bridge, 1988), Kärevere 3 (highway bridge, 1999), Kärkna (highway bridge, 2000), Raudtee (urban bridge, 1998), Kõrveküla (highway bridge, 2006) and Smuuli (urban bridge, 2007).

5.8.2 TALLIN'S CLIMATE

The cold temperate climate of Estonia is uniquely interesting as it provided an opportunity to examine the model's performance under extreme weather. Estonia's climate is characterized by breezy severe winters of temperatures as low as –5°C, while not exceeding 18°C during the relatively warmer months of May to September. Tallin's low precipitation of 673.3 mm arises during the seven (7) months of rainfall from October to April of the following year, while the five (5) months from May to September is a dry spell period of no rain. Being a town at the Baltic seacoast, Tallin has very high RH humidity ranging from 70% to 90%, giving an average of 83.2%. It may be recalled that such high RH >70% inhibits carbonation progression due to pore blockage by moisture presence (Ekolu, 2016).

5.8.3 COMPARISON OF THE MODEL'S PREDICTIONS WITH MEASURED NATURAL CARBONATION OF ESTONIAN BRIDGES

The *in situ* concretes cast were reportedly of moderate strengths ranging from 30 to 50 MPa. Unsurprisingly, the measured carbonation rates were very low, being between 0.2 to 1.5 mm/yr$^{0.5}$, an observation majorly attributed to the very high RH which inhibits carbonation progression. OPC concretes were assumed to have been used since no information had been reported on the cement types employed in mixtures. Some of the other parameters employed in the analysis were the *sheltered* outdoor exposure condition and estimated [CO_2] of 200 ppm, considering the city's coastal location at the Baltic seaside. It may be recalled that seas and oceans significantly absorb CO_2. It can be seen in Figure 5.10a, that the model made meaningful and realistic predictions. The residuals in Figure 5.10b also showed 'fanning-out' heteroskedasticity, which is consistent with findings for nearly all the prior carbonation analyses done in this chapter.

5.9 PREDICTION OF THE NATURAL CARBONATION OF 143 ITALIAN HIGHWAY STRUCTURES

5.9.1 DESCRIPTION OF THE EXISTING HIGHWAY CONCRETE STRUCTURES IN TURIN

Guiglia and Taliano (2013) investigated 143 new highway concrete structures all of ages no less than five (5) years old. They were wide-ranging types of highway structures comprising bridges, flyovers, underpasses, tunnels, viaducts and retaining walls, altogether covering a highway distance of 135 km. Concrete cores of 100 mm diameter extracted from the various structures, were tested to determine compressive strengths and carbonation depth values. Compressive strengths of the *in situ*

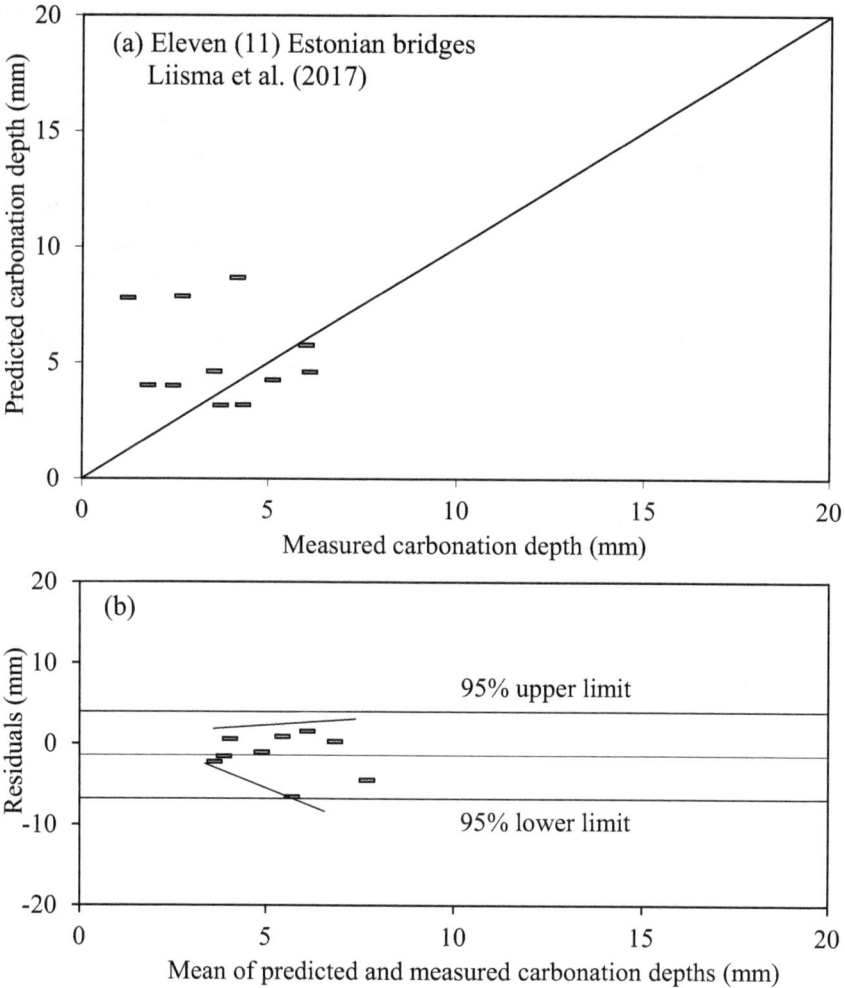

FIGURE 5.10   Eleven (11) highway and urban bridges in Tallin, Estonia (a) predicted versus measured carbonation depths and (b) residuals.

concretes were reportedly between 20 to 50 MPa. The cement types employed in the concrete mixtures were CEM I, CEM II/A-L and CEM II/B-L. The structures were grouped based on RH values of their site locations comprising Area I of 64% RH, Area II of 67% RH and Area III of 75% RH. The present carbonation analysis was limited to data of exposure Class XC3 specifically for *sheltered* structural elements, being mostly abutments, tunnels and piers.

5.9.2   TURIN'S CLIMATE

The weather pattern of Turin, Italy is a mild temperate climate somehow similar to that of Lugano, Switzerland (Section 5.6.2). However, Turin's annual precipitation of 284 mm is much lower than the 705.5 mm of Lugano. The climate in Turin comprises an annual RH of 74.3% and average temperature of 13.7°C.

5.9.3 COMPARISON OF THE MODEL'S PREDICTIONS WITH MEASURED NATURAL CARBONATION OF ITALIAN HIGHWAY STRUCTURES

Considering the wide range of the different highway structures investigated, the concretes utilized among various structural elements covered all strength grades comprising the low strengths of 20 to 30 MPa, moderate strengths of 30 to 50 MPa and high strengths > 50 MPa. The core strength values reported in the data, were converted to the equivalent cube strengths (Section 3.4, Eq. 3.9a). The highest concrete strengths employed were those of some piers in the Area I structures, wherein 70 to 85 MPa concretes were used.

Figure 5.11a–c shows graphs comparing the model's predictions against actual measured values of natural carbonation in the highway concrete structures of Area I,

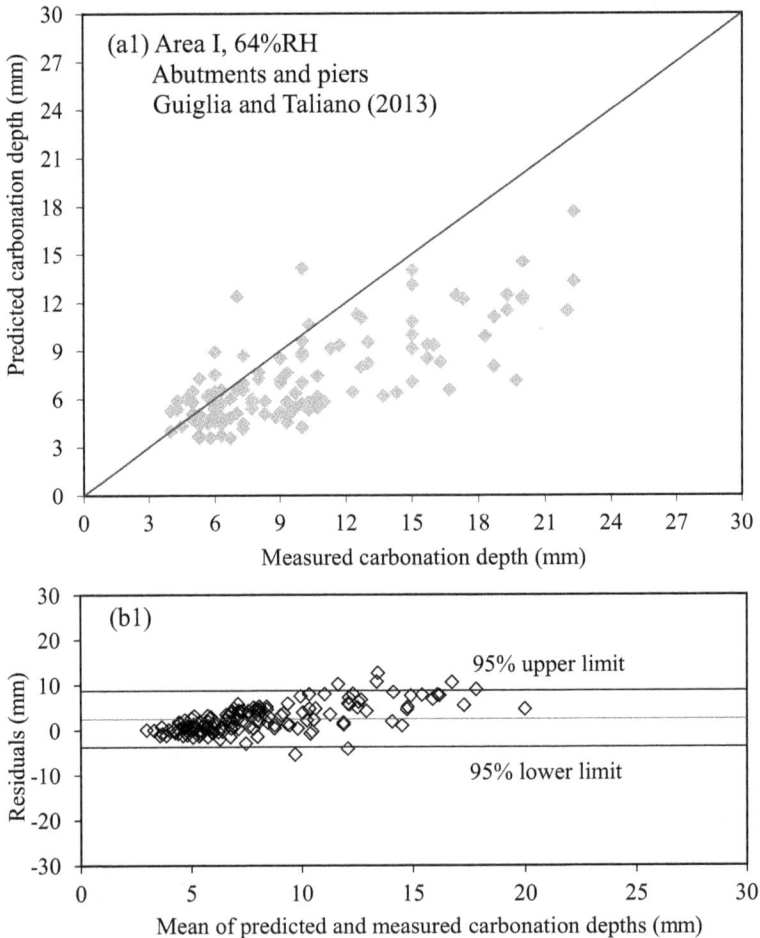

FIGURE 5.11a Area I highway structures in Turin, Italy (a) predicted versus measured carbonation depths and (b) residuals.

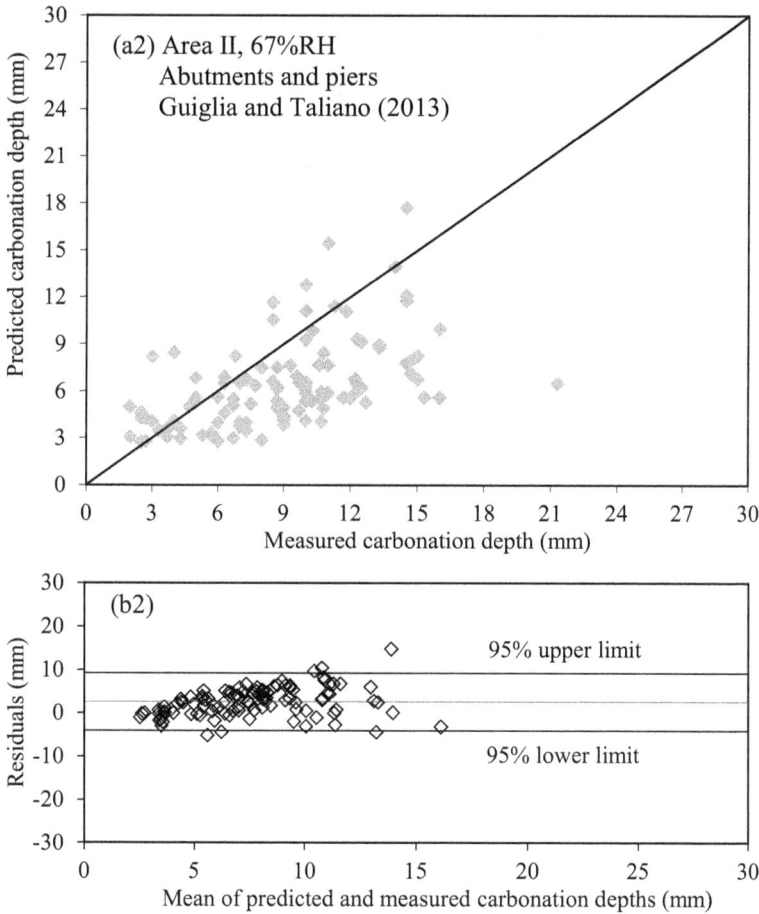

FIGURE 5.11b Area II highway structures in Turin, Italy (a) predicted versus measured carbonation depths and (b) residuals.

Area II and Area III, respectively. Altogether 345 data points of *sheltered* structural elements were plotted. Evidently, the model's prediction of natural carbonation in all the 143 highway structures, is realistic with values falling along the line of equality. Also evident is the typical 'fanning-out' heteroskedasticity seen in residuals of the structures located at Area I (Figure 5.11a (b1)), Area II (Figure 5.11b (b2)) and Area III (Figure 5.11c (b3)), respectively. The observed good accuracy of the model is indicated by its acceptable CV values comprising 39.8%, 45.6% and 44.2% for the structures located in Area I, Area II and Area III, respectively.

5.10 ERROR STATISTICS FOR ALL THE EVALUATED STRUCTURES

In the foregone Sections 5.2 to 5.9, hundreds of data points were employed to validate the model's prediction of natural carbonation in 208 real-life concrete structures.

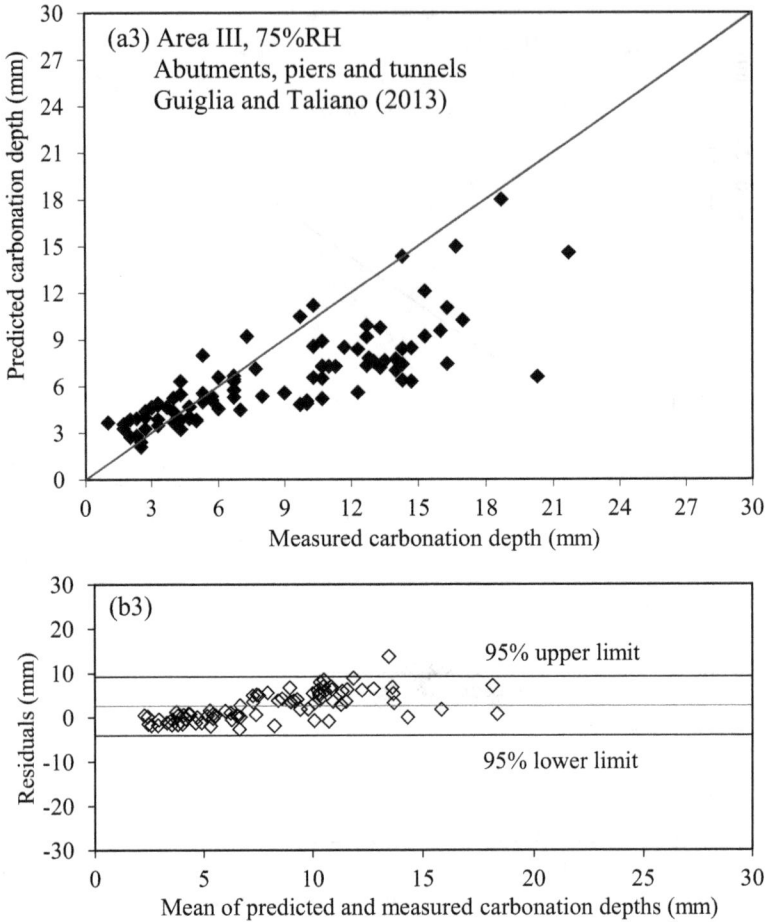

FIGURE 5.11c   Area III highway structures in Turin, Italy (a) predicted versus measured carbonation depths and (b) residuals.

The various existing reinforced concrete structures are located worldwide at different geographical regions (Figure 5.1) of widely different climates including the subtropical weather of Johannesburg (South Africa), the wet/dry tropical weather of Bhopal (India) and Taipei (Taiwan), the mild temperate climate of the Swiss Alpine region of Blenio in Lugano (Switzerland), the warm tropical equatorial climate of Brasilia (Brazil) and the cold temperate weather of Tallin (Estonia). The various real-life concrete structures evaluated were buildings, bridges and other types of highway structures. Both, new and old existing structures were among those evaluated, being of widely varied ages from one (1) year to 70 years at the time of testing.

Bazant and Baweja (1995b) employed an extensive RILEM databank to generate error values of code-type creep and shrinkage prediction models, as shown in Table 5.1. It can be seen in this table that the models B3, ACI 209 and CEB-FIP gave typical CV values of 20% to 60%.

TABLE 5.1
Coefficient of variation of errors for code-type models of creep and shrinkage (Bazant and Baweja, 1995b).

Model	B3 (1995)	ACI 209 (1992)	CEB-FIP (1990)
Shrinkage	34	55	46
Basic creep	24	58	35
Drying creep	23	45	32

TABLE 5.2
Statistical error indicators for all the structures.

Data source	Country	No. of structures	Types of structure/s	MV/PV*	RMS	CV (%)
Lampacher (2000)	South Africa	15	Bridges and buildings, Johannesburg	1.02	0.78	24.0
Rizvi et al. (2017)	India	10	Concrete structures, Bhopal city	1.21	8.27	32.3
Liang et al. (2013)	Taipei	1	Chorng-chin viaduct, Taipei	0.78	13.59	56.1
Liang et al. (2013)	Taipei	1	Wann-fwu bridge, Taipei	0.90	11.54	63.6
Guiglia and Taliano (2013), Area I	Italy	54	Highway structures, Turin	1.30	3.77	39.8
Guiglia and Taliano (2013), Area II	Italy	47	Highway structures, Turin	1.34	4.09	45.6
Guiglia and Taliano (2013), Area III	Italy	61	Highway structures, Turin	1.25	3.82	44.2
Teruzzi and Cadoni (2003)	Switzerland	2	Three-storey building and a gymnasium	1.17	3.57	22.6
Figueiredo and Nepomuceno (2005)	Brazil	6	Reinforced concrete buildings	0.75	10.11	44.6
Liisma et al. (2017)	Estonia	11	Urban and highway bridges	1.17	2.19	58.4

Note: *MV is the mean measured value, PV is the mean predicted value, RMS is the root mean square of error, CV is the coefficient of variation of error.

The NCP model's prediction accuracy, was evaluated based on three (3) different statistical error indicators comprising: the mean of measured values (MV) to mean of predicted values (PV) that is, MV/PV ratio, the root mean square of error (RMS) and coefficient of variation of error (CV). Given in Table 5.2 are error values indicating accuracy of the NCP model when employed to predict natural carbonation of the real-life concrete structures analyzed in Sections 5.2 to 5.9. It can be seen in Table 5.2 that the MV/PV ratio falls between 0.8 and 1.3, giving an average of a perfect 1.0 for all the various structures. As such, the model's overall predictions were generally similar to measured values of natural carbonation of the real-life concrete structures. It can be deduced from Table 5.2 that the NCP model effectively gives CV values of 20% to 60% which are the same as those of the code-type models B3, ACI 209 and CEB-FIP (Table 5.1). These findings show that the NCP model has the same prediction accuracy as code-type models.

REFERENCES

ACI 209 (1992, 1997) *Prediction of creep, shrinkage and temperature effects in concrete structures*, American Concrete Institute (ACI) Committee 209, Subcommittee II. Report ACI 209 R92 (re-approved 1997).

Bazant Z.P. and Baweja S. (1995a) Justification and refinements of Model B3 for concrete creep and shrinkage, 1. Statistics and sensitivity, *Materials and Structures*, 28, 415–430.

Bazant Z.P. and Baweja S. (1995b) Creep and shrinkage prediction model for analysis and design of concrete structures: model B3. In: *Adam Neville symposium: creep and shrinkage – structural design effects*, ACI SP–194 (A. Al-Manaseer, Ed.), American Concrete Institute, Farmington Hills, Michigan, 2000, 1–83. Also described in ACI *Concrete International ACI 23*, January 2001, 38–39.

CEB-FIP Model Code (1990) *Design code 1994*, Thomas Telford, London.

Ekolu S.O. (2016) A review on effects of curing, sheltering, and CO_2 concentration upon natural carbonation of concrete, *Construction and Building Materials*, 127, 306–320.

Ekolu S.O. (2020) Model for natural carbonation prediction (NCP): practical application worldwide to real life functioning concrete structures, *Engineering Structures*, 224, 111126, http://doi.org/10.1016/j.engstruct.2020.111126

Figueiredo C.R. and Nepomuceno A.A. (2005) The carbonation of reinforced concrete buildings in Brazil. In: *Application of codes, design and regulations, proceedings of the international conference held at University of Dundee*, Scotland, UK, 5–7 July, 221–228, Thomas Telford, ISBN 9780727734037.

Gardner N. and Lockman M.J. (2001) GL-2000 design provisions for drying shrinkage and creep of normal strength concrete, *ACI Materials Journal*, 98, 159–167.

Guiglia M. and Taliano M. (2013) Comparison of carbonation depths measured on in-field exposed existing r.c. structures with predictions made using fib-Model Code 2010, *Cement and Concrete Composites*, 38, 92–108.

Lampacher B.J. (2000) Assessing the durability of concrete structures, *PhD Thesis*, School of Civil and Envir Engineering, University of the Witwatersrand, 212p.

Liang M.T., Huang R. and Fang S.A. (2013) Carbonation service life prediction of existing concrete viaduct/bridge using time-dependent reliability analysis, *Journal of Marine Science and Technology*, 20(1), 94–104.

Lifecon (2003) *Deliverable D 3.2 service life models: life cycle management of concrete infrastructures for improved sustainability*, Final Report by Dipl.-Ing. Sascha Lay, Technical Research Centre of Finland (VTT), 169p.

Liisma E., Sein S. and Järvpõld M. (2017) The influence of carbonation process on concrete bridges and durability in Estonian practice, *IOP Conference Series: Materials Science and Engineering*, 251, 012072.

Rizvi S.S., Akhtar S. and Verma S.K. (2017) Carbonation induced deterioration of concrete structures, *The Indian Concrete Journal*, September, 6p.

Teruzzi T. and Cadoni E. (2003) Application of a life-time management method on existing concrete structures of 25 scholastic facilities by probabilistic estimation of the residual service life. In: *Integrated lifetime engineering of buildings and civil infrastructures ilcdes 2003*, 1–3 December, Kuopio, Finland, 1001–1006.

6 Probabilistic service life design and analysis

6.1 CORROSION RISK EXPOSURE CATEGORIES

Steel corrosion in concrete structures is significantly influenced by climate. Being an atmospheric agent, $[CO_2]$ is sufficiently present globally to induce carbonation of reinforced concrete in all human-inhabited regions comprising the temperate, tropical and subtropical climate zones. EN 206 (2000) divides corrosion risk exposures into three (3) categories comprising: (i) corrosion induced by carbonation, (ii) corrosion induced by chlorides other than those from seawater and mainly involve de-icer salts, and (iii) corrosion induced by chlorides from seawater.

SABS 0100–2 (1992, Section 3.3.3) states that marine atmospheric conditions exist from the sea coastline up to 15 km inland. The foregoing implies that inland reinforced concrete structures located beyond 15 km from the seacoast, are considered safe from exposure to chloride attack, as also reported by Alao et al. (2014). Table 6.1 gives the scope of risk categories suggested for application of the NCP model. Evidently, the model is applicable to all climates (except the polar zones) for above-ground concrete structures. For coastal structures, both carbonation and chloride attack occur but the latter is more critical. Indeed, chloride attack on steel reinforcement is relatively more severe than carbonation-induced attack (Costa and Appleton, 2001; Kulkarni, 2009). As such, service life design of coastal structures typically focusses on the chloride attack mechanism, while inland reinforced concrete structures under the risk category XC3, are designed against carbonation-induced attack. For coastal structures, it is presumed that design against chloride attack simultaneously protects the structure against carbonation-induced attack. That is, chloride attack is the control damage mechanism for coastal structures and no independent design against carbonation may presumably be necessary for such structures built of reinforced concrete (Kulkarni, 2009; Berke, 2006).

6.2 CLIMATE AND ENVIRONMENTAL CONDITIONS OF THE AFRICAN CONTINENT

6.2.1 CLIMATE ZONES

Africa has a long coastline of 30,500 km owing to the continent's large land mass, being the second largest in the world. The African continent is surrounded by three (3) water bodies comprising the Indian Ocean on the east and south, Atlantic Ocean on the west and south and the Mediterranean Sea in the north. Being salt-laden, seas and oceans are the main chloride sources for reinforcement corrosion at coastal

DOI: 10.1201/9781003645399-6

TABLE 6.1
Risk categories for application of the NCP model.

Climate		Location of concrete structures	Mechanism	Applicability
Tropical, subtropical	Inland	Beyond 15 km from seacoast	Carbonation	NCP model
	Coastal	Within 15 km from seacoast	Carbonation Chlorides from sea water	NCP model
Temperate	Inland	Beyond 15 km from seacoast	Carbonation	NCP model
	Coastal	Within 15 km from seacoast	Carbonation Chlorides from seawater	NCP model

regions of the African continent. The severity of chloride-induced corrosion attack largely depends on distance of the structure from the coastline. Within 0 to 100 m from the coastline, also referred to as the marine zone, structures are considered to be in contact with water and chloride-induced corrosion is designated to be *extremely severe*. Within 100 m to 15 km from the coastline, structures are exposed to *severe* category of chloride-induced corrosion. At the inland zone which is typically located at the distance greater than 15 km from the coastline, chloride attack is non-existent and corrosion is considered to be *moderate* (Haque, 2006; Kulkarni, 2009).

Figure 6.1 is a map of African continent showing the inland and coastal zones along with associated prevailing carbonation and/or chloride attack mechanisms that typically occur in these regions. Also included in Figure 6.1 are Africa's various climate zones comprising the tropical, subtropical, temperate and arid zones. Evidently, a majority of large cities and urban infrastructures are predominantly located in the four (4) climate zones.

6.2.2 ENVIRONMENTAL FACTORS

In this section, major cities have been selected from each climate zone across Africa, for comparison of their environmental parameters of interest. Five (5) cities were randomly selected from each of the main inland zones comprising the tropical, subtropical and arid climates. Also selected were six (6) coastal cities of different climates. Figure 6.2 shows locations of the selected cities across different climate zones of the continent. In the figure, a number is assigned to each city, as also given in Table 6.2 along with annual average values of environmental parameters comprising precipitation, RH and temperature.

Figures 6.3 to 6.5 show climate graphs giving the monthly precipitation, RH and temperature for the selected inland cities located in the tropical, subtropical and arid zones, respectively. The climate graphs were plotted using data drawn from https://en.climate-data.org. Similarly given in Figure 6.6 are the environmental parameters for selected coastal cities located in various climate zones.

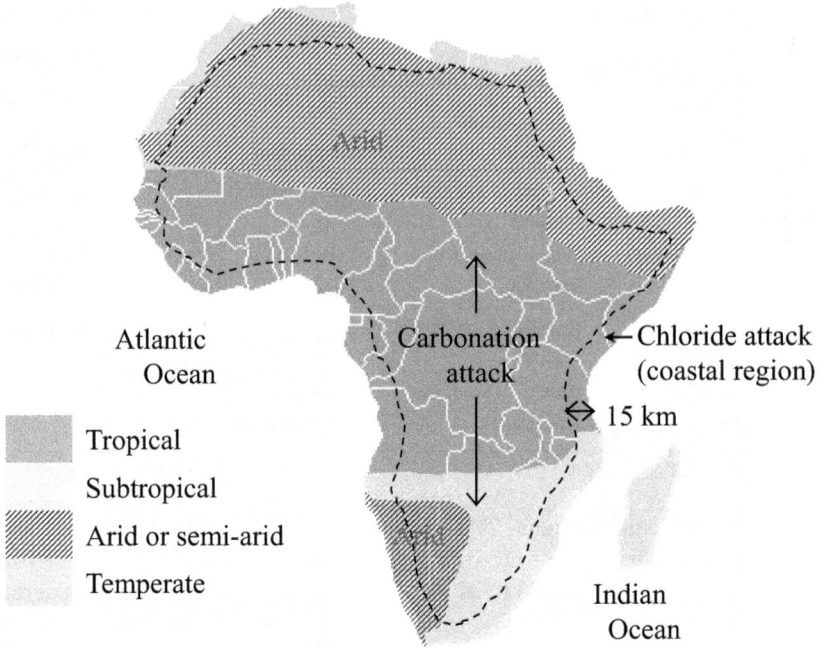

FIGURE 6.1 Inland and coastal zones of the African continent.

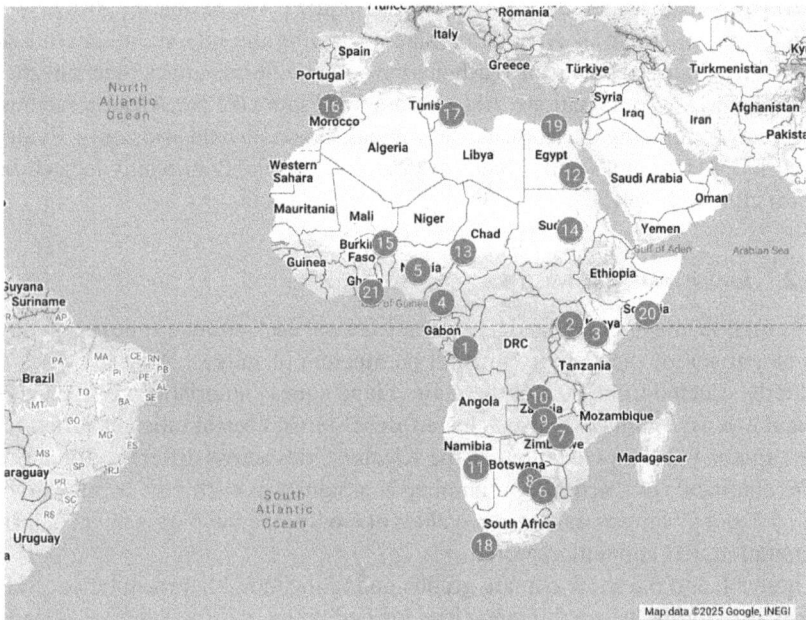

FIGURE 6.2 African cities selected across different climate zones of the continent: 1–5 are in tropical zones, 6–10 are in subtropical zones, 11–15 are in arid or semi-arid zones and 16–21 are coastal cities (https://www.google.com/maps).

TABLE 6.2
Climate conditions of cities across Africa.

Climate	City and location			Annual average values of		
	Name and country	No	Location	Precipitation (mm)	Relative humidity (%)	Temperature (°C)
Tropical	Kinshasa, DRC	1	Inland	1095	77.3	25.5
	Kampala, Uganda	2	Inland	1747	82.0	21.4
	Nairobi, Kenya	3	Inland	674	64.0	18.8
	Yaoundé, Cameroon	4	Inland	1727	82.9	23.0
	Abuja, Nigeria	5	Inland	1469	60.2	26.0
Subtropical	Johannesburg, S. Africa	6	Inland	784	59.3	16.3
	Harare, Zimbabwe	7	Inland	824	58.9	18.6
	Gaborone, Botswana	8	Inland	510	45.8	20.7
	Lusaka, Zambia	9	Inland	970	61.0	20.4
	Lubumbashi, DRC	10	Inland	1098	59.5	21.4
Semi-arid /arid	Windhoek, Namibia	11	Inland	534	38.2	18.5
	Aswan, Egypt	12	Inland	0	22.8	26.6
	N'Djamena, Chad	13	Inland	424	35.3	28.9
	Khartoum, Sudan	14	Inland	70	25.2	29.9
	Niamey, Niger	15	Inland	283	35.5	29.1
Temperate	Rabat, Morocco	16	Coastal	442	74.6	18.1
Arid	Tripoli, Libya	17	Coastal	354	59.6	20.7
Subtropical	Cape Town, S. Africa	18	Coastal	621	75.2	16.4
Arid	Alexandria, Egypt	19	Coastal	784	65.3	20.8
Semi-arid	Mogadishu, Somalia	20	Coastal	530	74.5	26.4
Tropical	Accra, Ghana	21	Coastal	746	81.4	26.4

Source: (data from https://en.climate-data.org).

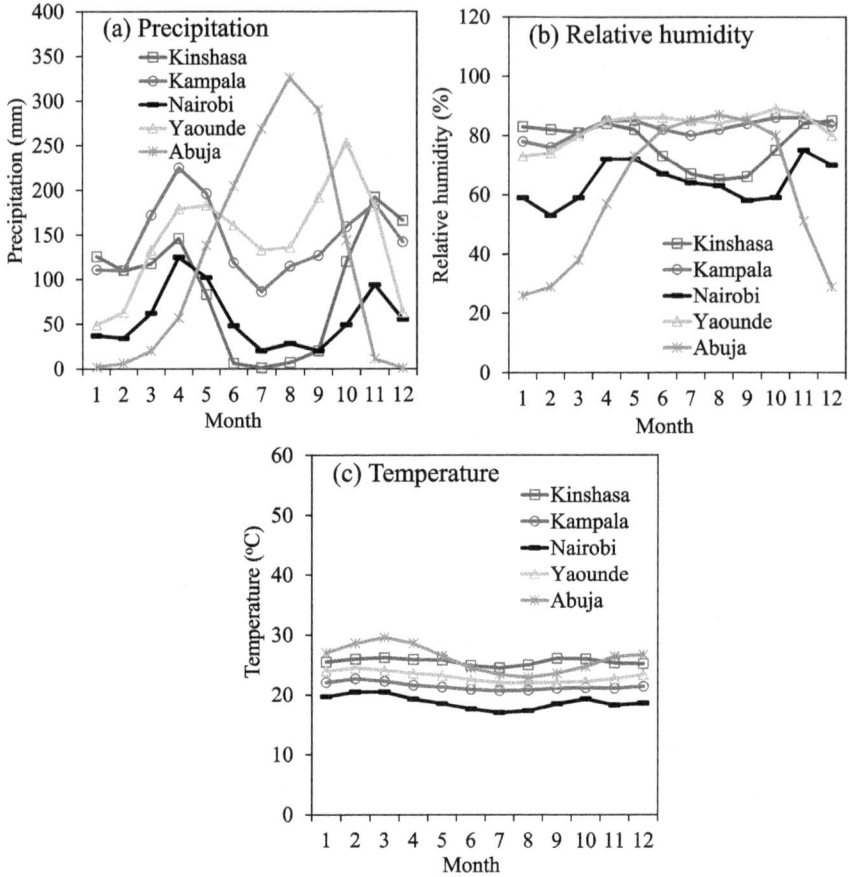

FIGURE 6.3 Tropical climate of randomly selected inland cities across Africa.

Precipitation and RH are the most distinctive factors that differentiate the various climate zones. Tropical zones have high precipitation of generally 1000 to 2000 mm, along with high RH levels of 60% to 85%. It can be seen in Figure 6.3a that the tropical climate has a double peak or single peak wet season. The cities in East and Central Africa comprising Kampala, Kinshasa, Nairobi and Yaoundé, experience a two-peak wet season running from March to May and from September to November. Moreover, outside the peak months of precipitation, some cities such as Kampala and Yaoundé still receive significant rainfall while for other cities such as Nairobi and Kinshasa, little to no rain falls during the off-peak period. Westwards of the tropical zone (Figure 6.3a), West African cities such as Abuja and Bamako, experience a single peak of heavy rainfall over the six-month period from May to October.

The subtropical zone exhibits a characteristic single wet season of rainfall from November to March of the following year, as seen in Figure 6.4a. As expected, rainfall in arid or semi-arid climates is little to none. Cities including Windhoek,

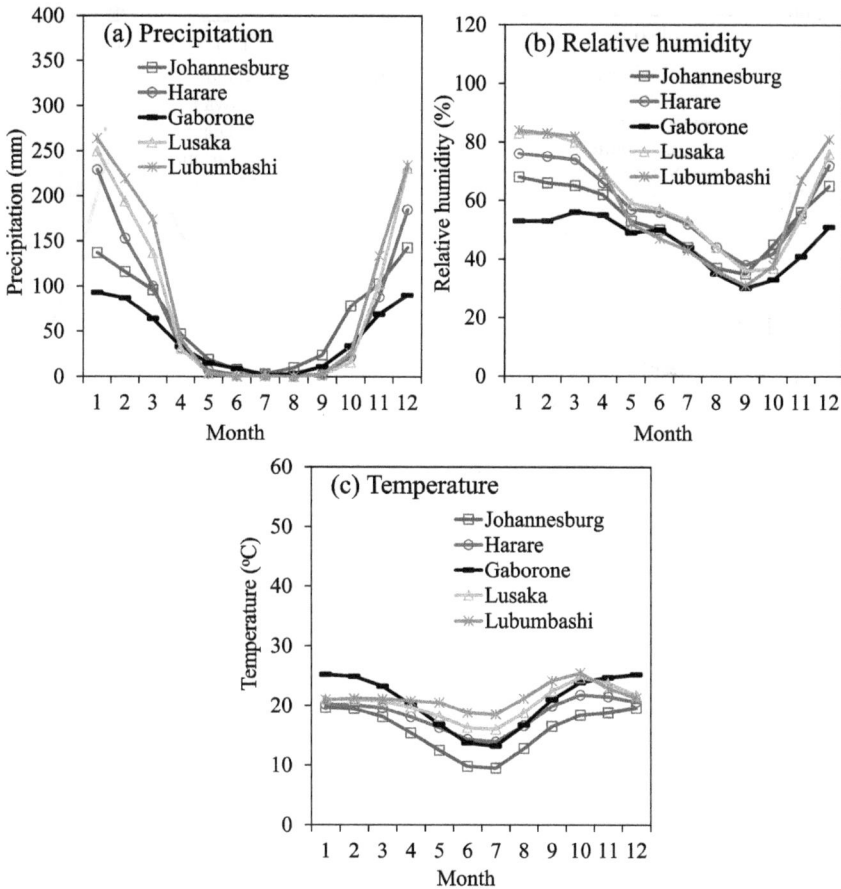

FIGURE 6.4 Subtropical climate of randomly selected inland cities across Africa.

N'Djamena and Niamey that are located in the semi-arid climate, receive little rain-fall while others such as Aswan and Khartoum that are in the arid region, hardly receive any rain at all (Figure 6.5a).

In the inland regions, RH levels and fluctuations generally follow the rainfall pattern. Hence inland cities that are located in the tropical climate zone have high RH of 60% to 85% (Figure 6.3b), followed by 50% to 60% for cities in subtropical regions (Figure 6.4b), 30% to 50% for those in semi-arid regions and below 30% for cities located strictly in arid climates (Figure 6.5b). In contrast to inland cities, it is evident that RH levels of coastal cities do not follow the rainfall or precipitation pattern of their climates. Instead, RH in coastal cities is consistently high, owing to proximity to the sea or ocean. The RH levels at coastal zones vary from 60% to 70% for cities such as Tripoli and Alexandria that are in arid or semi-arid regions, to RH high levels of 70% to 80% for coastal urban centres located in tropical, subtropical

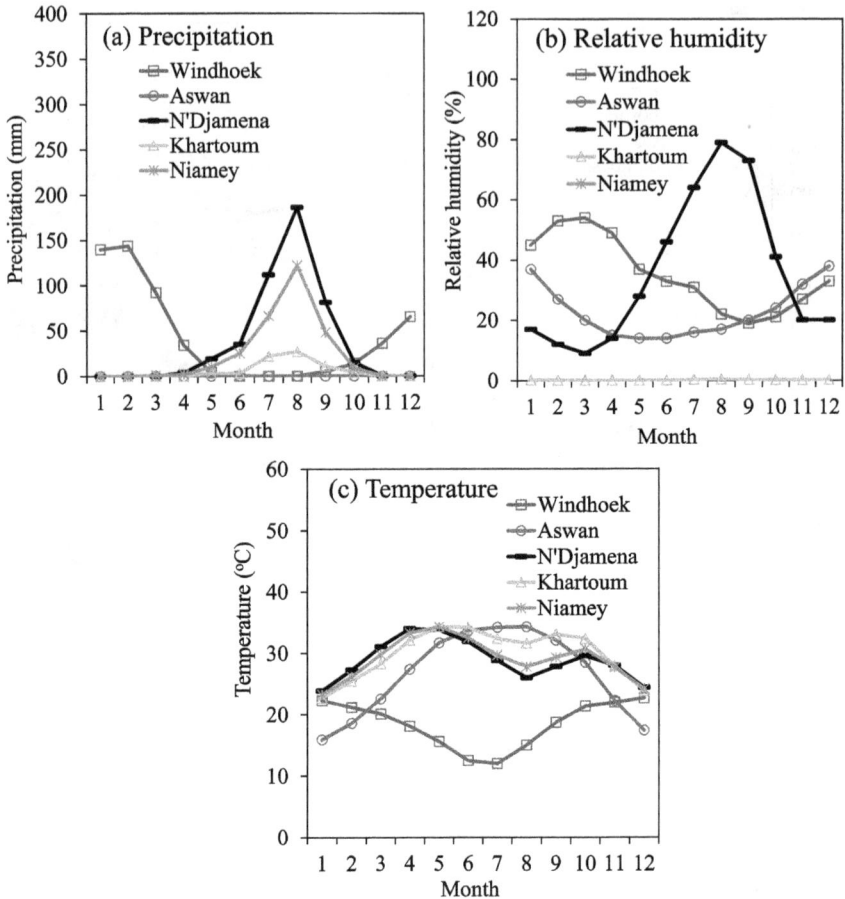

FIGURE 6.5   Arid and semi-arid climates of randomly selected inland cities across Africa.

or temperate climates. Clearly, coastal cities have characteristically and steadily high RH levels of 60% to 80% throughout the year as seen in Figure 6.6b, while inland cities have fluctuating RH levels as per their rainfall patterns (Figures 6.3a, b to 6.5a, b). Table 6.3 gives the proposed reasonable classification values representative of environmental conditions prevailing at the different climate zones.

Generally, no severe winters occur among the African climate zones, hence weather conditions across the continent are generally warm to hot with average annual temperatures ranging from 15 to 30°C. Major differences arise from the extent of temperature fluctuations. The tropical climate is characterized by steady monthly temperatures of 20 to 30°C throughout the year, as seen in Figure 6.3c. For example, the monthly average temperature of Kampala city is 21.4 ± 1.0°C throughout the year, while that of Abuja is consistently 26.0 ± 3.4°C. In contrast, the non-tropical climates are characterized by large temperature fluctuations during various seasonal changes throughout the year, as seen in Figures 6.4c, 6.5c and 6.6c.

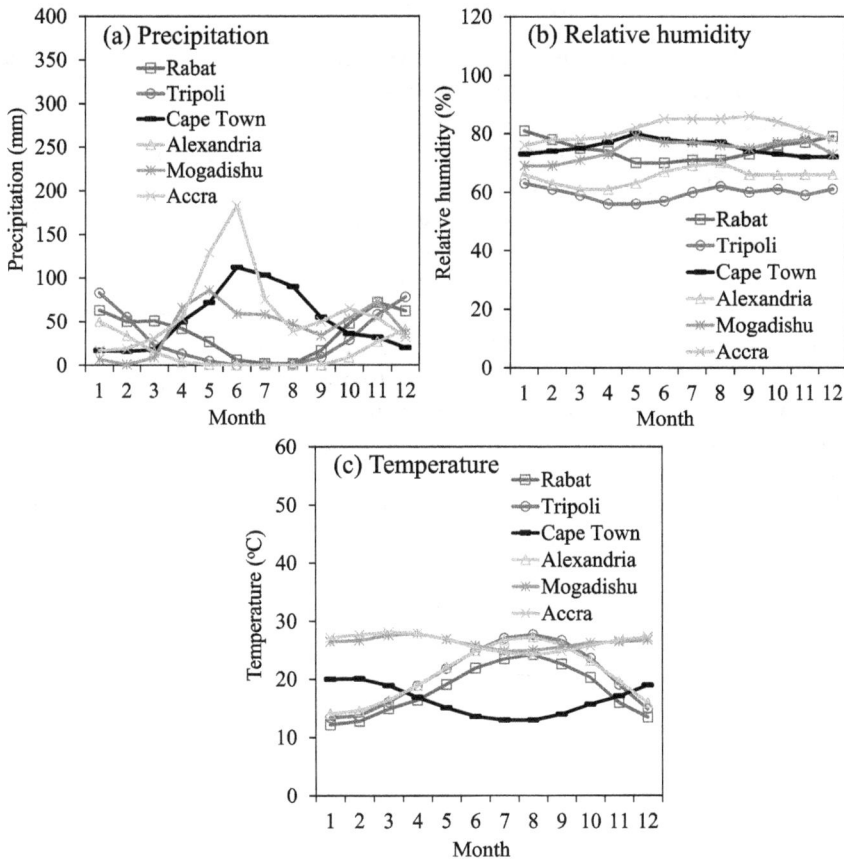

FIGURE 6.6 Climates of various African coastal cities.

TABLE 6.3
Proposed classification of African climate zones.

Location	Climate type	Annual precipitation (mm)		Annual relative humidity (%)		Annual temperature (°C)
Inland	Tropical	High	>1000–2000	High	>60–85	Steady 18–25
	Subtropical	Moderate	>500–1000	Moderate	50–60	16–20
	Semi-arid	Low	>100–500	Low	30–49	25–30
	Arid	Very low	0–100	Very low	<30	25–30
Coastal	Variable			Steadily high	60–85	Variable

6.2.3 INFLUENCE OF CLIMATE ON CONCRETE CARBONATION ACROSS AFRICAN CONTINENT

In Africa and other developing countries, the concrete strengths predominantly used in reinforced concrete structures are 20 to 30 MPa. To illustrate the influence of climate on carbonation progression across Africa, the concrete strength of 25 MPa was chosen for use in the analysis. The NCP model (Chapter 3) has been employed to predict the 50-year carbonation progression while keeping [CO_2] constant at 400 ppm breached in 2014 (https://research.noaa.gov).

This analysis compares the carbonation effects of different climate types, considering all the variables including temperature influence. Figure 6.7 shows the predicted carbonation depths in the selected cities located at the various climate zones. The analysis was based on 25 MPa concrete made with OPC. It can be seen in Figure 6.7a, b and d that while carbonation progression in the tropical zone is significantly high, it is similar or slightly lower than that of the subtropical climate or coastal exposure. Expectedly, concrete carbonation in the arid or semi-arid regions is nil or negligible (Figure 6.7c).

RH is the crucial factor responsible for the foregoing observed differences in carbonation responses of concrete structures located in the different climate regions. The characteristically high RH level of the tropical climate reduces carbonation progression, as pores are blocked by moisture presence. At the other extreme, RH in the semi-arid or arid zones is too low, thus CO_2 remains mostly in gaseous state due to absence of sufficient moisture needed to support the carbonation reaction (Chapter 2). The highest carbonation occurs in the subtropical region wherein the typical RH

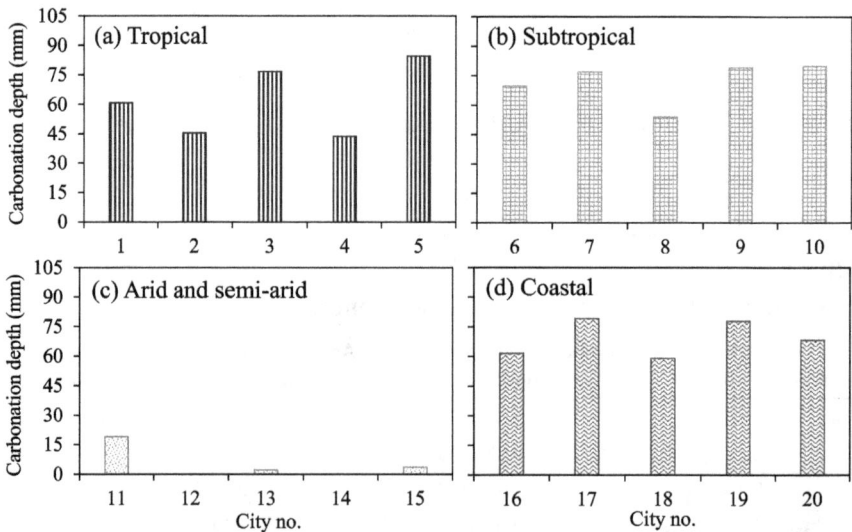

FIGURE 6.7 The 50-year carbonation progression in 25 MPa concretes of structures across African cities of various climate zones (city names are given in Table 6.2).

levels of 50% to 60% are ideal for maximum carbonation. Generally, however, all the three (3) climates comprising the tropical, subtropical and coastal areas, are high carbonation zones. Eventually, the local climate conditions control the extent of carbonation attack. For example, arid zones that are close to water bodies can also be carbonation prone areas owing to elevated RH levels of the microclimates.

6.3   SERVICE LIFE DESIGN AND ANALYSIS

This section focusses on application of the NCP model to service life design (SLD) and analysis of concrete structures, based on the stochastic approach. Theoretical underpinnings of the applicative stochastic method are discussed in Chapter 2. The SLD methodology involves two (2) main steps consisting of:

1. Carbonation prediction using the NCP model, which requires employment of six (6) independent variables comprising (Chapter 4): two (2) materials parameter inputs of compressive strength and cement type, along with four (4) environmental parameters of RH, $[CO_2]$, sheltering and temperature.
2. Reliability index analysis or Monte Carlo simulation, which additionally involves employment of the cover to steel reinforcement and coefficient of variation (CoV) values or standard deviations of independent variables.

The end-of-service life (ESL) is more appropriately evaluated based on failure probability. For structures exposed to chloride attack, 10% failure probability is recognized as the ESL criterion. Indeed, in Norway, 10% failure probability is specified in the standard as the serviceability limit state (Gjorv, 2009; Standard Norway NS 3490, 2004). This stringent criterion is adopted as a safety measure for protection of structures against severity of the chloride attack mechanism. Carbonation-induced corrosion, however, is less severe than chloride attack. As such, it is generally recognized that the failure criterion for carbonation can be less stringent than that of chloride attack.

In the literatures, different ESL criteria have been considered for structures subject to carbonation-induced corrosion. Teruzzi and Cadoni (2003) employed ESL criterion of 50% failure probability when evaluating the carbonation of Swiss concrete structures (Chapter 5, Section 5.6). Liang et al. (2012) suggested that the point at which 'rust-expansion-cracks' occur, can be taken as the ESL criteria. This suggested criterion would typically be earmarked by an observation of brown staining at the concrete surface. It is typically the case, that stains appear at the concrete surface after the carbonation front has reached and/or gone past the level of steel reinforcement. Consistent with the foregoing, it is possible that perhaps 30% to 50% of the carbonation front would have reached or gone past the level of steel reinforcement, before stains begin to appear at the concrete surface. However, there is presently no generally recognized or standardized ESL criterion for carbonation-induced corrosion, in which case further research is needed.

Failure probability, which is the statistical basis of prescribing the ESL, is determined from reliability index analysis (Section 2.5.2) or from Monte Carlo simulation (Section 2.5.4). Reliability index (β) is calculated using Equation (2.9) from which the probability distribution function is determined (Sarja and Vesikari, 1996). Equation (2.9) is also written more appropriately as Equation (6.1).

$$\beta(t) = \frac{R - \mu[S,t]}{\sqrt{(CoV \cdot \mu[S,t])^2 + (CoV \cdot R)^2\,]}} \tag{6.1}$$

where R is resistance, S is loading, μ is the mean of, CoV is the coefficient of variation, β is the value along x-axis of NORMDIST curve (0,1).

It is well-established from the literatures (Sarja and Vesikari, 1996; Ekolu, 2010) that the typical CoV values of carbonation depth and cover depth are 0.4 and 0.2, respectively.

Subsequently given are examples of service life analysis for concrete structures selected from worldwide locations comprising Johannesburg (South Africa), Taipei (Taiwan) and Bhopal (India). In the examples, *in situ* strength data were measured directly from the existing or real-life concrete structures that had undergone natural carbonation during service over the years, hence the models Equations (3.7a, b) given in Chapter 3 are the appropriate equations to employ in the calculations. Table 6.4 summarizes the analyses parameter variables employed, which also account for urban locations of the various structures. The rationale for employment of the stated parameters, are discussed under each subsequent subsection.

TABLE 6.4
Random variables and parameter inputs employed in service life analyses.

Random variable	Existing reinforced concrete structures		
	Johannesburg structures, S. Africa	Wann-fwu bridge, Taipei, Taiwan	Bhopal structures, India
RH (%)	59.3	70.2	41.8
Sheltering	Sheltered	Sheltered	Sheltered
Temperature (°C)	16.3	26.0	28.3
CO_2 (ppm)	325	370	400
Strength, f_{chn} (MPa)	35.3 (for 1Yshl), 38.7 (for 15Bshl), 43.8 (for 12D2shl), 52.7 (for 10Pshl), 61.2 (for 6Eshl)	38.3 (average), 39.9 (deck), 37.9 (beam). (11 data points for all elements)	20 (for S5shl), 25 (for S17shl), 28 (for S16shl), 40 (for S10shl)
(*cem, g*) for CEM I	(1000, −1.5)	(1000, −1.5)	(1000, −1.5)
Cover (mm)	40	44 (average value), 40 (for deck), 50 (for beam)	35 (for S5shl), 45 (for S17shl), 32 (for S16shl), 35 (S10shl)
CoV for cover depth	0.2	0.2	0.2
CoV for $d_{c(t)}$	0.4	0.4	0.4

6.3.1   RELIABILITY INDEX-BASED SERVICE LIFE ANALYSIS
OF JOHANNESBURG STRUCTURES

The description of Johannesburg structures along with their data, are given in Chapter 5, Section 5.3. It may be recalled that the NCP model gave realistic prediction of the ongoing carbonation progression in the existing fifteen (15) bridges and buildings, which in turn validated the model's veracity for employment in service life analysis. Five (5) Johannesburg structures built with concretes of strengths 35, 40, 50 and 60 MPa (Table 6.5), were selected for service life analysis.

All the structures were of *sheltered* exposure. Apart from 1Yshl (Table 6.5) which was built in 1922, all the other Johannesburg structures were built between 1962 and 1973, during which period the $[CO_2]$ varied between 318 ppm and 330 pm, hence the average $[CO_2]$ of 325 ppm was used in service life analysis. OPC was the common cement type used in all the structures, which is consistent with concrete technology of the time. The cover depth of 40 mm typically specified for bridge construction in South Africa, was adopted for use in the analysis.

Figure 6.8 shows probabilistic service life prediction for the Johannesburg structures. Also included in the figure are lines indicating the 10% and 30% failure probability criteria. The calculated ESL values for each of the structures, are given in Table 6.5 for both failure criteria. For the *1Yshl* structure, it can be seen that the ESL of 25 years based on 10% failure probability, is unrealistic as the structure

---

TABLE 6.5
End-of-service life values along with field inspection observations.

Structure	Age during testing in 1992 (years)	Age in 2022 (years)	End-of-service life, ESL (years)		Field observations
			10% F(P)	30% F(P)	
1Yshl	70	100	25	65	The structure was repaired after about 70 years and demolished around the year 2015, then replaced due to change of priorities by the user or owner.
15Bshl	19	49	40	>100	By the time of this analysis in 2022, the structure was functional with no known major repairs done.
12D2shl	20	50	70	>100	By the time of this analysis in 2022, the structure was functional with no known major repairs done.
10Pshl	22	52	>100	>100	By the time of this analysis in 2022, the structure was functional with no known major repairs done.
6Eshl	24	54	>100	>100	By the time of this analysis in 2022, the structure was functional with no known major repairs done.

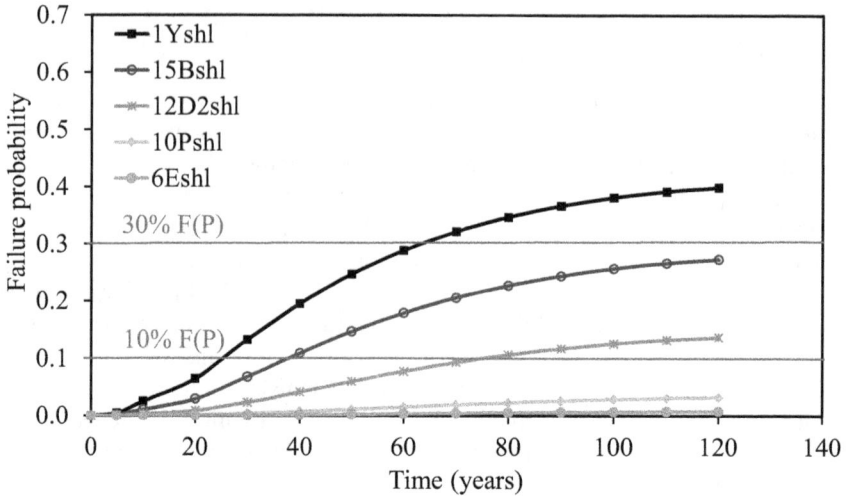

FIGURE 6.8 Probabilistic prediction for service lifespans of the Johannesburg structures: 30% F(P) is 30% failure probability.

continued to function until demolition at the age of about 94 years. The repairs done on the structure around 1992 at the age of about 70 years, coincides closely with the ESL of 65 years estimated based on 30% failure probability. Also, the structure *15Bshl* that was 50 years in 2022 would have existed past the 40-year ESL based on 10% failure probability. The inspection done in 2022 by the book author, shows that neither repairs nor any related works had been conducted on *15Bshl*. Rather, the structure remained functional without deterioration concerns, which again is consistent with the estimation based on 30% probability, giving ESL of at least 100 years. Concerning the other three (3) structures comprising *12D2shl*, *10Pshl* and *6Eshl*, it is too early to assess accuracies of the 10% and 30% failure probability criteria. Nonetheless, the foregoing observations suggest that employment of 10% failure probability criterion, seems unrealistically stringent for carbonation-induced corrosion, while the 30% failure probability criterion may be reasonable to consider as the basis for ESL estimation.

6.3.2 RELIABILITY INDEX-BASED SERVICE LIFE ANALYSIS OF WANN-FWU BRIDGE

Chapter 5, Section 5.5 provides the description details of Wann-fwu bridge located in Taipei. A total of eleven (11) data points were reported comprising three (3) measurement locations done on the deck, and seven (7) results from the beam and from the abutment wall. The concrete cover of the bridge deck was 40 mm, while that of the cap beam and abutments was 50 mm. The [CO_2] of 370 ppm was used in the analysis. The other input parameter variables used, are given in Table 6.4.

Service life prediction for Wann-fwu bridge is shown in Figure 6.9. The analysis was done collectively for average carbonation response, as well as separately for structural elements comprising the bridge deck and the cap beam. Based on 30% failure probability, the cap beam performance is higher than that of the bridge deck,

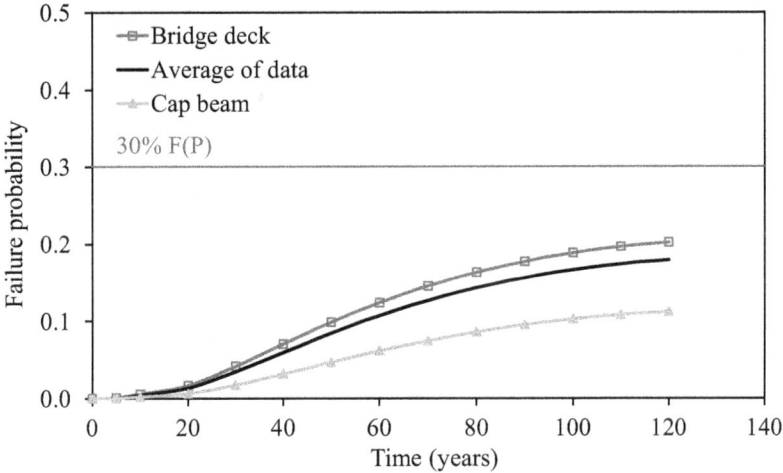

FIGURE 6.9 Probabilistic service life prediction for Wann-fwu bridge.

but both give ESL values greater than 120 years. Evidently, the service life of the bridge is at least 100 years, mainly owing to the high concrete strength and large cover used in construction, along with high RH of the climate (Table 6.4).

6.3.3 RELIABILITY INDEX-BASED SERVICE LIFE ANALYSIS OF BHOPAL STRUCTURES

Bhopal structures that were analysed for service lifespans, are described in Chapter 5, Section 5.4. Four (4) representative Bhopal structures were selected consisting of *S5shl*, *S17shl*, *16Shl* and *S10shl*, built with concretes of strengths/cover depths comprising 20 MPa/35 mm, 25 MPa/45 mm, 28 MPa/32 mm and 40 MPa/35 mm, respectively. It may be noted that each of the selected structures had a different concrete strength along with a different cover depth. Being generally young structures built at end of last century or within the first quarter of the 21st century, the [CO_2] of 400 ppm was used in the analysis. OPC was taken to be the cement type used in the concretes.

Figure 6.10 shows the service life prediction functions for the Bhopal structures. A short ESL of 25 years was obtained for the low strength 20 MPa concrete structure, *S5shl*. The moderate 32 mm cover and moderate 28 MPa concrete strength for *S16shl*, are responsible for the observed good ESL of 70 years. The structures *S17shl* and *S10shl* that had moderate or high strengths and cover depths, both exhibited ESL values exceeding 100 years.

6.4 RELIABILITY INDEX ANALYSIS METHODOLOGY

6.4.1 STEP-BY-STEP PROCEDURE

When implementing the NCP model in accordance with the procedure described in Chapter 3 Section 3.6, correction factors are calculated using the model's Equations (3.2) to (3.9). The calculated factors are then entered into the model's Equation (3.1) to

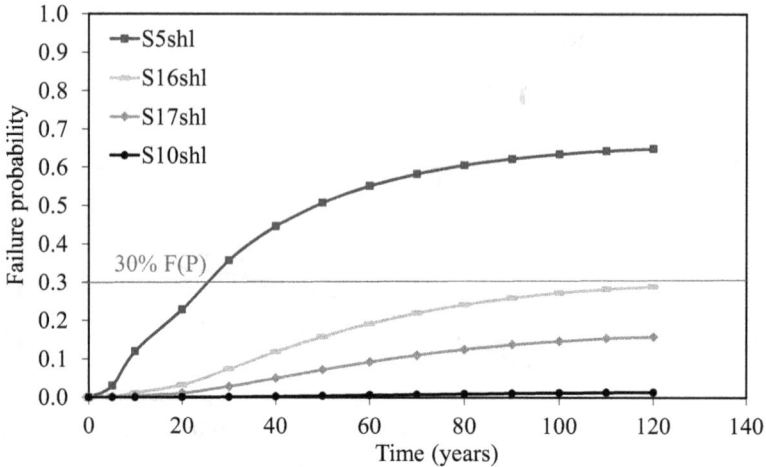

FIGURE 6.10   Probabilistic prediction for service lifespans of Bhopal structures.

obtain a unique carbonation depth-time relationship for every different data point. As such, the NCP model determines the carbonation rate coefficient (K_c) as a dynamic parameter calculated as $K_c = e_h.e_s.e_c.e_t.$ *cem* $(F_{ct})^g$, which differs for each concrete mixture exposed to different environmental parameters and conditions. Evidently for a given concrete mixture, the model flexibly makes time-based adjustments to the carbonation-time relationship under different independent variables. The step-by-step procedure of conducting reliability index analysis calculations, is subsequently given.

Step 1. Check the condition that concrete strength, $f_c > 20$ MPa is satisfied.

Step 2. Determine the strength growth function (F_{ct}) (model's Eqs. 3.6a to 3.7b).

Step 3. Calculate the carbonation depth function ($d_{c,t}$) (model's Eqs. 3.1 to 3.5; 3.8), by following the detailed procedure given in Chapter 3, Section 3.6.

Step 4. Determine the reliability index (β) function (Eq. 6.1).

Step 5. Calculate β for the varied ages spaced at 5-year intervals from t = 1 to 120 years, such as, at t =1, 5, 10, 15, 20. . .120 years. To calculate β for each age, follow Steps 2 to 4.

Step 6. Use the EXCEL function to calculate failure probability as F(P) = 1-NORMDIST(β;0;1;1). Alternatively, standard normal distribution tables can be used.

Step 7. Repeat Steps 5 and 6 for all the ages. Plot a graph of age on the x-axis against F(P) on the y-axis.

6.4.2   Example of reliability index analysis calculations

Consider an existing concrete structure constructed using CEM I concrete, exposed to weather conditions comprising the annual average RH of 59.3%, annual temperature of 16.3°C, [CO_2] of 400 ppm and *sheltered* from rain. The *in situ* concrete strength and cover depth are 35 MPa and 40 mm, respectively.

Step 1. $f_c = 35 > 20$ MPa OK.

Step 2. For ages < 15 years, use the model's Eq. 3.7a.

For ages ≥ 15 years, use the model's Eq. 3.7b.

Step 3. The model's input parameters are: $e_h = 1.010$ (model's Eq. 3.2), $e_s = 1.000$ (model's Eq. 3.3), $e_c = 1.122$ (model's Eq. 3.4), $e_t = 0.970$ (model's Eq. 3.5) and (cem, g) = (1000, −1.5) (model's Eq. 3.8a, b).

$$\text{Therefore, } S_t = d_{c,t} = (1.010)(1.000)(1.122)(0.970) \cdot 1000 \left(F_{c,t}\right)^{-1.5} \cdot \sqrt{t}$$

$$= 1099.2 \left(F_{c,t}\right)^{-1.5} \times \sqrt{t}$$

Step 4. Reliability index, $\beta = \dfrac{40 - S_t}{\sqrt{\left(0.4 \times S_t\right)^2 + \left(0.2 \times 40\right)^2}}$

Step 5. Consider ages t = 1, 5, 10, 15, 20. . .120 years and calculate β for each age as shown below.

At t = 30 years,

$$F_{c,t} = \dfrac{30}{0.15(30) + \left(0.95 - \dfrac{30^{0.6}}{50}\right) \cdot 30} .35 = 36.995$$

$$S_t = 1099.2(36.995)^{-1.5} \times \sqrt{30} = 26.756$$

$$\beta = \dfrac{40 - 26.756}{\sqrt{\left(0.4 \times 26.756\right)^2 + \left(0.2 \times 40\right)^2}} = 0.991$$

Step 6. Calculate F(P) = 1-NORMDIST(0.991;0;1;1) = 0.16.

Step 7. Repeat Steps 5 and 6 for all the ages, to generate the probability distribution function given in Table 6.6 and plotted in Figure 6.11.

6.5   EXAMPLE OF MONTE CARLO SIMULATION

Step 1 to 3. Follow the same steps given in Section 6.4.2.

Step 4. For each random variable, assign CoV and type of probability distribution as given in Table 6.7 and employ the NORM.INV function to run the inbuilt RAND() function.

TABLE 6.6

Probability distribution function.

t (years)	1	5	10	20	30	. . .	90	100	110	120
F(P)	0.00	0.01	0.03	0.08	0.16	. . .	0.40	0.41	0.43	0.43

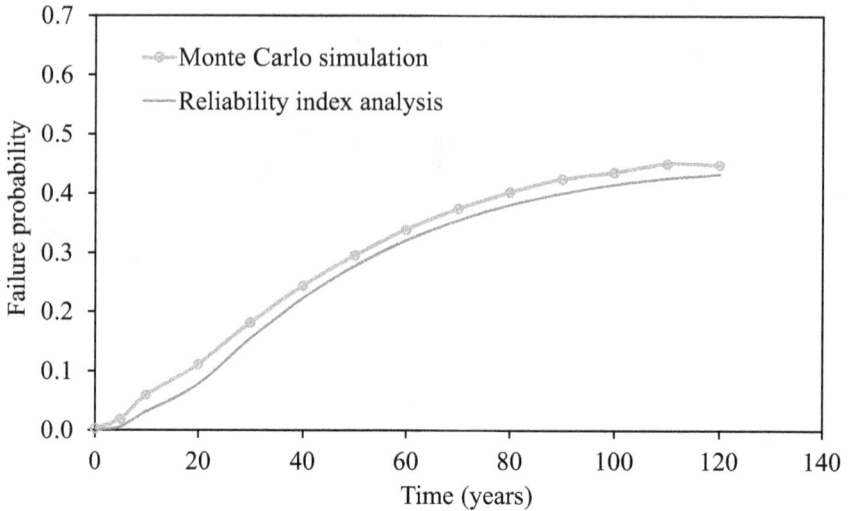

FIGURE 6.11 Comparison of probability distribution functions generated based on reliability index analysis and Monte Carlo simulation.

TABLE 6.7
Random variables calculated using Equations (3.1) to (3.8) of the NCP model and assigned types of distributions.

Variable	Mean	Standard deviation	Distribution
e_h (59.3% RH)	1.01	0.09696	Normal
e_s (sheltered)	1		Constant
e_{co} (400 ppm)	1.122	0.14025	Normal
e_t (16.3°C)	0.97	0.04074	Normal
cem	1000		Constant
g (CEM I)	−1.5		Constant
f_c (MPa)	35	7	Normal
Cover (mm)	40	8	Normal

Step 5. At each age t = 1, 5, 10, 15, 20 . . . 120 years, compute the Outcome = $c - d_{c,t}(X)$ (Eq. 2.10) for N = 30000 simulations.

Step 6. At each age, compute F(P) using Equation (2.11) to generate the probability distribution function.

Figure 6.11 compares the probability distribution function generated using Monte Carlo simulation with that obtained in Section 6.4.2 based on reliability index analysis. It can be seen that although there can be some differences, the functions generated using both methods are generally similar and consistent with realistic performance of the model.

REFERENCES

Alao O.O., Alexander M. and Beushausen H. (2014) Understanding the influence of marine microclimates on the durability performance of RC structures. In: *Proceedings of the first international conference on construction materials and structures* (S.O. Ekolu, et al., Eds.), Johannesburg, South Africa, 1060–1067.

Berke N.S. (2006) Corrosion of reinforcing steel. In: *Significance of tests and properties of concrete and concrete-making materials*, ASTM STP 169D, 164–173.

Costa A. and Appleton J. (2001) Carbonation and chloride penetration in a marine environment, *Concrete Science and Engineering*, 3, 242–249.

Ekolu S.O. (2010) Model validation and characteristics of the service life of Johannesburg concrete structures. In: *Proceedings of the national symposium on concrete for a sustainable environment*, Concrete Society of Southern Africa, 3–4 August, Kempton Park, Johannesburg, Gauteng, 30–39.

EN 206 (2000) *Concrete – part 1: specification, performance, production and conformity*, European Committee for Standardization, CEN, Management Centre: Rue de Stassart, 36 B-1050, Brussels, Belgium.

Gjorv O.E. (2009) *Durability design of concrete structures in severe environments*, Taylor and Francis, UK, 19p.

Haque M.N. (2006) African concrete code-design for durability. In: *2nd African concrete code symposium*, University of Stellenbosch, Western Cape, South Africa, 15p.

Kulkarni V.R. (2009) Exposure classes for designing durable concrete, *The Indian Concrete Journal*, March 23–43.

Liang M.-T., Chang H.-T. and Yeh C.-J. (2012) Rust-expansion-crack service life prediction of existing reinforced concrete bridge/viaduct using time-dependent reliability analysis, *Journal of Marine Science and Technology*, 20(4), 397–409.

SABS 0100–2 (1992) *Code of practice for the structural use of concrete, part 2: materials*, South African Bureau of Standards, Pretoria.

Sarja A. and Vesikari E. (1996) *Durability design of concrete structures*, RILEM Report 14 (A. Sarja and E. Vesikari, Eds.), E & FN Spon, UK, 93p.

Standard Norway NS 3490 (2004) *Design of structures: requirements to reliability*, Standard Norway, Oslo (In Norwegian).

Teruzzi T. and Cadoni E. (2003) Application of a life-time management method on existing concrete structures of 25 scholastic facilities by probabilistic estimation of the residual service life. In: *Integrated lifetime engineering of buildings and civil infrastructures ilcdes 2003*, 1–3 December, Kuopio, Finland, 1001–1006.

Index

accelerated carbonation 16–17, 19, 26, 30–31, 35

accuracy of the model 68, 81; *see also* prediction accuracy

actual measured values 7, 47, 65, 70, 80

adaptability 27, 44

affordability 6; *see also* cost-effectiveness

African nations 6; *see also* sub-Saharan Africa

age of the structure 69

aggressive agents 4, 19

ambient temperature 32, 34, 38

analytical mathematical methods 64

anthropogenic CO_2 11, 18; *see also* carbon-dioxide

applicability 7, 28, 33, 35, 44, 87

arid 11, 87–95; *see also* climate zones

Arrhenius 19, 31

ASTM C595 57

ASTM Type I 57

atmospheric CO_2 15, 40, 52, 69, 86; *see also* carbon-dioxide

Austin 43–44, 46–48, 51, 61

Bhopal structures 68, 70, 96, 99–100

blast-furnace slag 16, 32, 35, 43, 53–54, 72; *see also* ground granulated blast-furnace slag

blended cement/s 28, 53, 57, 61, 66

Blenio 65, 72, 75, 82

BOUTEK 51–52, 61

Bouzoubaa 57, 61

Brasilia 65, 76–77, 82

Brazilian buildings 77

bridges 1, 65–67, 69–70, 72, 77–78, 82–83, 97

Bucher 58, 61

buildings 65–66, 69, 72, 74–77, 82–83, 97

burnt shale 32, 35

calcite 11–12, 14

calcium hydroxide 11, 40

calcium silicate hydrate 11, 40

carbonation: attack 3, 11, 17–18, 22, 88, 95; coefficient 14, 58, 68; conductance 35; data 26, 38, 42–43, 46, 64; effect/s 16, 17, 94; front 13, 22, 95; measurements 54; mechanism 7, 11; phenomenon 3, 14–16, 27, 64; rates 14, 57–58, 68, 70, 72, 74, 78; reaction/s 11–14, 15–16, 17, 19, 94; research 14–15, 18, 42; testing 16, 26, 67; theory 16; *see also* phenolphthalein indicator

carbonation modelling 3–6, 18–19, 40, 42; *see also* models

carbonation prediction 17, 19, 54, 74, 84, 95

carbonation progression 2, 4, 16–17, 23, 33, 36, 42–43, 49–50, 65, 77–78, 94, 97

carbonation: attack 3, 11, 17–18, 22, 88, 95; coefficient 14, 58, 68; conductance 35; data 26, 38, 42–43, 46, 64; effect/s 16, 17, 94; front 13, 22, 95; measurements 54; mechanism 7, 11; phenomenon 3, 14–16, 27, 64; rates 14, 57–58, 68, 70, 72, 74, 78; reaction/s 11–14, 15–16, 17, 19, 94; research 14–15, 18, 42; testing 16, 26, 67; *see also* phenolphthalein indicator; theory 16

carbon-dioxide 2, 12: anthropogenic 11, 18; atmospheric CO_2 15, 40, 52, 69, 86; *see also* CO_2 concentration

cement: blends 60–61; composites 27, 43, 48, 50, 60–61; designations 32, 46, 53; factors 32; 35; hydration 11, 28, 40; type/s 27–28, 32, 35, 38, 40–41, 43, 52–55, 58, 72, 78–79, 95, 97, 99; varieties 46

cementitious: materials 7, 16, 35; matrix 12; system/s 18, 40–41 42, 59–60; *see also* supplementary cementitious materials

Changsha 43–44, 46–48, 51, 61

Chennai 43–48, 51, 61

chloride-laden environment 71; *see also* marine environment

chloride/s: attack 1, 3, 11, 17, 19–20, 40, 86–87, 95; conductivity 65; induced corrosion 1; 20, 87; critical 20; ingress 52, 64

Chorng-ching viaduct 70–73

climate/s: change 1–2, 6, 8–11, 18, 27, 62; factors 46; regions 3, 17, 27, 31, 65, 94; resilience 2; variations 44, 48, 50, 63; worldwide 46

climate zones: arid 87, 89; equatorial 44, 46, 48, 76, 82; global 3, 6–7, 9–10, 39, 42, 46, 61–62; subtropical 3, 52, 68, 82, 86–91, 93–94; temperate 1, 3, 11, 31, 44, 46, 48, 57, 59, 78–79, 82, 86–89; tropical 1, 3, 11, 17, 31, 44, 48, 69–71, 76, 82, 86–94

clinker cement 43

CO_2 concentration 2, 13, 16–17, 19, 31, 33–34, 38, 40–41, 46, 52, 70, 72

CO_2 emissions 2, 6–7, 46, 52

coastal 1, 3, 11, 52, 61, 78, 86–89, 91, 93–94: cities 3, 87–88, 91; regions 11; structures 1, 86

code-type models 56, 61, 70, 76, 84; *see also* models

coefficient of variation 5, 50–51, 56–57, 83, 95–96

coefficient of variation of error 50–51, 83

compositional factors 40–41

compressive strength/s 5–6, 12, 16–17, 28–29, 40–41, 54, 59, 65, 69, 71, 77–78, 95, *see also* concrete strength/s

computational intelligence 42

For Product Safety Concerns and Information please contact our EU
representative GPSR@taylorandfrancis.com
Taylor & Francis Verlag GmbH, Kaufingerstraße 24, 80331 München, Germany

www.ingramcontent.com/pod-product-compliance
Lightning Source LLC
Chambersburg PA
CBHW032003180326
41458CB00006B/1717